Mary Camilla Foster Hall Wood

Santa Barbara as It is

Mary Camilla Foster Hall Wood

Santa Barbara as It is

ISBN/EAN: 9783337379681

Printed in Europe, USA, Canada, Australia, Japan

Cover: Foto ©berggeist007 / pixelio.de

More available books at **www.hansebooks.com**

Santa Barbara As It Is.

TOPOGRAPHY,

CLIMATE,

RESOURCES,

AND

OBJECTS OF INTEREST.

Published by the
INDEPENDENT PUBLISHING CO.,
Santa Barbara, Cal.

INDEX.

Santa Barbara As It Is.

INTRODUCTORY.

IN presenting this brief description to the public, our aim is not so much to lavish poetic praises upon an "Earthly Paradise," as to gather into convenient and consistent shape the practical details of soil, agriculture, climate and society, which will answer the questions naturally asked by tourists or intending settlers.

Although much has been written about Santa Barbara, good, bad and indifferent, we feel that there is still something to be said; that the new material, together with a condensed statement of all that has been previously noted, may yet find interested readers. Still, we enter with diffidence upon the task, aiming only to present to our readers a resumé of Santa Barbara's natural gifts and advantages, her hotels, schools and societies, her fauna and flora, her mineral and agricultural wealth.

We have drawn largely upon whatever was of definite value in other publications. We are also greatly indebted to certain gentlemen to whom credit will be given in the proper place, whose names are of recognized authority upon the subjects whereof they individually treat.

Within the limited space allowed, our only object is to present such a picture of Santa Barbara that the most enthusiastic reader shall not be tempted to exclaim, how delightful!—but rather, how true, how plain, how accurate!

The exact truth about Santa Barbara is enough, and more than enough. Exaggerations injure us in large measure. Leading the stranger to expect but a single ray of sunshine more than he finds, renders him discontented with the most genuine of fair facts. But when all is said, the real glory of the land, its gentle and gracious atmosphere, cannot be explained to a northern ear. One must live in it, be thoroughly steeped in it, to love it as its native races do. The old saying, applied to many other pleasant places on the earth's surface, is particularly true of this: "Whoever has lived in it a year, is sure to come back to it at last."

"It is not your mountains, or magical chain
Of islands dim purple, nor even the sea,
With gay racing billows by day, and by night
His monotone chant to uncomforted souls;
Not these, but the spirit of these, but the breath,
The reviving, the incomprehensible air,
That we breathe in and live in and love till we die!"

M. C. F. Hall-Wood.

SANTA BARBARA COUNTY,

FROM 1542 TO 1884.

A BRIEF HISTORICAL SKETCH.

CHAPTER I.

THE earliest authentic record of Santa Barbara is found in the narrative of Juan Rodriguez Cabrillo, a Portuguese navigator in the service of Spain, who, on the 27th day of June, 1542, just fifty years after the landing of Christobal Colon on the island of San Salvador, sailed northward from the western coast of Mexico, with two vessels, the San Salvador and the Victoria. He passed through the channel of Santa Barbara, observing upon the islands a dense population, "men and women of fair complexion." He claims to have spent months among the natives of what is now Santa Barbara county, and notes the names of forty populous villages on the mainland. After voyaging up the coast, Cabrillo returned, landing upon San Miguel island, where he died and was buried. All traces of his last resting-place have long since been obliterated by the shifting sands. Sixty years later, Sebastian Vizcaino, commanding a Spanish fleet, re-discovered the channel and named it "Canal de Santa Barbara," presumably because it was reached upon the 4th of December.

Three hundred years ago, these valleys supported a swarming population. Its soil freely supplied nuts and wild fruits. Game abounded on the land and fish in the channel. The simple and indolent aborigines wore no clothing worth mentioning. All their necessities were furnished by a fertile soil with little labor, and their days were passed in calm security and animal comfort. At long intervals strange ships appeared in the roadstead. Strange men landed, procured supplies and sailed away. Gradually, it became known to the brawny rovers who sailed under the victorious banners of Spain, that California was not an island and not a peninsula, that the great river to the north did not connect with the Atlantic Ocean, and that neither the Isles of the Amazons nor the seven cities of Cibola lay in this direction.

The strangers continued to come, encroaching more and more upon the territory of the simple inhabitants, until this primitive race slowly faded away before the resistless rovers. Their day had reached its twilight one hundred years ago, when the Catholic religion and the Spanish arms had gained a firm foothold upon the Pacific Coast.

The first efforts of Christian civilization in this county were made by the Franciscan friars. The Mission of Santa Barbara was established in 1786, the eleventh in point of time among the Missions of California; that of San Diego, in 1769, being the first. After a few years of toil and privation the Holy Fathers began to grow in power and opulence. They were the temporal as well as the spiritual lords of the land. They cultivated, by the hands of their Indian converts, the olive, the fig and the vine, enjoying all the luxuries of a genial climate and a generous soil that this great and costless labor could produce. The whole race of natives were in fact serfs whose toil was repaid by spiritual blessings. Until 1830 the Fathers held undisputed sway. Missions were established all down the coast, making a close cordon from San Francisco to San Diego. In this county the Mission la Purisima was established in 1787; that of Santa Ynez in 1804.

From the year 1800 Anno Domini to 1822, (when the Spanish dominion was finally overthrown in Mexico,) were the halcyon days of the Mission system. The Fathers dwelt in patriarchal state; with regal revenues and powers. In 1825, it is recorded that the Mission of Santa Barbara possessed 75,000 cattle, 5,000 tame horses and mules, and 40,000 sheep.

But this also passed away. Mexico was in a chaotic, transitional state,—hurrying from revolution to revolution. Every new government sent new plunderers to loot the wealthy Missions. Finally, in 1833, the Mexican Government took possession of the Missions, and the greater number of them were allowed to fall into ruins. Only the Missions of Santa Ynez and Santa Barbara were retained by the Franciscans. After the priestly dominion was followed by the military and civil rule of Mexico, now independent of Spain, Santa Barbara became a place of importance. Here certain Governors of the province resided, and here were held Departmental Assemblies, until the time of Pio Pico, in 1846, the last of the old time Governors. In 1841 Los Angeles was the largest town in the Californias, and Santa Barbara and San Francisco were of equal size.

In 1846, the question of annexing California to England was discussed and decided negatively in an Assembly convened in Santa Barbara.

Then in 1849, came the grand event of the Pacific Coast; the discovery of gold. In the mad rush for sudden wealth, which drew the adventurous of all classes into its seething vortex, Southern California was almost abandoned. New centres of wealth and population were formed as if by magic. For twenty years this portion of the State lay in its golden sunlight, smiling and unknown. Rumors then began to spread of an unparalleled climate below Point Concepcion. Straggling invalids made their way here and wrote back ecstatic accounts of balmy, delicious airs and everlasting sunshine. In 1872, along with a crowd of health seekers from all parts of the United States, came Charles Nordhoff, an experienced journalist, whose enthusiasm and ardent praises awakened an unbounded interest in all parts of our common country, spreading even to Europe. The inhabitants of Santa Barbara awoke one morning and found their cherished valley famous!

From 1872 to 1875, nothing less than insanity ruled. Prices of property advanced, doubled, quadrupled! It was in those days that the country and climate were puffed out of all reason. Consumptives in the last stages of the disease thronged the streets and the hotels. Even private houses were filled with ghastly guests who ought never to have left their homes. The inevitable crash came, following on the heels of the Eastern panic,—and this was a ruined community.

In time, business recovered its tone, slowly but surely. The old enthusiasm gave way to a more even growth, and to-day Santa Barbara can boast of a solid and steady prosperity beyond the reach of any ordinary panic.

CHAPTER II.

TOPOGRAPHY.

THE County of Santa Barbara, roughly stated, is in shape a parallelogram, bordered on the south and west by the Pacific ocean, on the north by San Luis Obispo County and on the east by Ventura County. The southern side of the parallelogram, nearly sixty miles in length, begins at Point Concepcion, where the coast suddenly trends to the eastward, and ends at Point Rincon, when it again assumes a southeasterly course. An extra depression in the coast line just where the city of Santa Barbara is situated, curves in so deeply that the frontage of the city itself has decidedly a southeastern aspect, its streets thereby running diagonally across from the cardinal points of the compass, from northeast to southwest and from northwest to southeast.

From Point Concepcion to the northern edge of the county is a distance of about thirty-five miles, as the crow flies; but on this western border of our parallelogram, Point Arguello, Point Purisima and Point Sal reach out into the Pacific, more or less boldly.

A large portion of Santa Barbara County is covered with mountain ranges. Parallel with the southern shore and the Santa Barbara Channel with its outlying islands, rise the rugged heights of Santa Ynez. These traverse the county from west to east, a solid range, from 3,000 to 4,000 feet high. Beyond the Santa Ynez, the San Rafael range lies, running in a northwesterly direction, almost at right angles. Beyond these again to the northeast, the irregular peaks of the Coast Range combine to form a wild melange of mountains, only broken by the valleys of the Cuyama. The greater part of this northeastern quarter is set down upon official maps as unsurveyed Government land. Yet among these inaccessible heights are many narrow valleys, cañons and cañadas, watered by mountain springs, pleasant and fertile; occupied by adventurous settlers and traversed by intrepid hunters or the still more intrepid prospector, in search of mineral wealth.

Between the Santa Ynez and the sea lies the unparalleled valley of Santa Barbara. It is about forty-five miles in length, averaging perhaps three miles in width. It contains a matter of some 86,400 acres of land. It is this belt which has become noted the world over for its semi-tropical climate and phenomenal fertility.

The valley of Santa Barbara proper, extending from the Rincon to Point Concepcion, comprises the valley of the Carpinteria, from the Rincon to a small spur of the Santa Ynez called Ortega hill, a distance of about nine miles; the Montecito, from Ortega hill to the city limits, some six miles; the city of Santa Barbara itself, occupying a space fully two miles square and still reaching out on all sides; and eight miles beyond, the small village of Goleta. Beyond Goleta, still following the broad avenue, the great ranchos of Hollister and Cooper, Los Dos Pueblos, Nuestra Señora del Refugio, and other ranchos brings the traveler to the Gaviota Pass, and a few miles beyond the Pass, Point Concepcion. Here the Santa Ynez range runs bodily into the Pacific Ocean, and the valley of Santa Barbara comes to an abrupt termination.

Beyond the Santa Ynez range, beginning near the eastern edge of the county, slowly opens between the Santa Ynez and San Rafael ranges the lovely valley of the Santa Ynez. The Santa Ynez river, running almost due west from its mountain source, parallel with the islands, the channel, Santa Barbara valley and the Santa Ynez mountains, waters a wide extent of agricultural lands, and near its mouth runs through the broad valley of Lompoc, emptying at last into the

Pacific, between Point Concepcion and Point Purisima. The towns of Lompoc and Santa Ynez are contained in this valley.

The next valley, Los Alamos, is watered by an arroyo of the same name, which rises in the San Rafael mountains and empties into the Pacific between Point Purisima and Point Sal, though sometimes sinking into the ground and becoming invisible. Los Alamos is a long, narrow valley, scarcely more than two miles wide in its broadest part, and contains but one town, that of Los Alamos.

The great Santa Maria valley lies upon the northern limits of the county. It is said to contain at least 65,000 acres of tillable land. This is certainly underestimated. The Santa Maria river, forming the northern boundary line of the county, is a continuation of the Cuyama, which waters the valleys of Cuyama in the northeast, the only open portion of the piled up mountain ranges of that quarter. The valley of Santa Maria proper contains the towns of Santa Maria, formerly called Central City, and Guadalupe, near the mouth of the river.

Three of the channel islands also belong to Santa Barbara County. San Miguel, the most western of the group, is about 2½ miles in width, and 7½ in length. Santa Rosa contains 53,000 acres and rises to a height of 1,172 feet. Santa Cruz, lying almost opposite the city of Santa Barbara, at a distance of twenty-five miles, contains 52,760 acres, and attains a height of 1,700 feet.

The county is still divided into the ranchos of the Mexican period. As a matter of special interest, and with the kind permission of Joseph J. Perkins, Esq., of this city, we append hereto the very valuable table compiled by him; showing at a glance these rancho divisions, with their acreage, etc. All lands not enrolled among these ranchos are either pueblo lands or unsurveyed Government lands. The city of Santa Barbara and the suburbs of Montecito and Carpinteria are built upon pueblo lands.

Names of Ranchos.	District	*Acres.*	*Cattle.*	*Sheep.*	*Horses.*
Lompoc and Mision Vieja de la Purisima	Lompoc District.	46,499 04-100	950	2,000	500
La Espada y El Cojo. } Forming Punta de la Concepcion	Lompoc District.	24,992 04-100	1,400		40
W. ½ Nuestra Senora del Refugio	Lompoc District.	13,265	Stock in Julian.	cluded	with San
San Julian	Lompoc District.	48,221	575	64,703	7
Canada de Sal Si Puedes	Lompoc District.	6,656 21-100	Stock in Julian.	cluded	with San
Santa Rosa	Lompoc District.	16,525 55-100	78	17,000	20
Santa Rita	Lompoc District.	13,316 05-100	50		200
Mision de la Purisima (Malo)	Lompoc District.	34 012 56-100		8,000	50
South ½ Jesus Maria	Lompoc District.	20,000	200	4,000	20
		233,487 45-100	3,253	95,703	837
*Cuyama Nos. 1 and 2	Santa Maria District.	71,020 75-100	3,000		
*Suey	Santa Maria District.	48,834			40
Tepusquet	Santa Maria District.	8,900	200	250	200
Sisquoc	Santa Maria District.	35,485 90-100	660	4,200	19
Tinaquaic	Santa Maria District.	8,874 60-100	1,200	500	60
*Punta de la Laguna	Santa Maria District.	26,648 42-100	300	4,000	300
*Guadalupe	Santa Maria District.	43,681 85-100	1,500	5,000	300
		243,445 52-100	6,860	13,950	919
North ½ Jesus Maria	Los Alamos District.	22,184 93-100	300	6,000	20
La Laguna	Los Alamos District.	48,703 91-100	250	10,000	100
Los Alámos	Los Alamos District.	48,803 38-100	500	25,000	300
Todos Santos	Los Alamos District.	20,772 17-100	200	3,000	50
Casmalia	Los Alamos District.	8,841 21-100	150	6,000	25
		149,305 60-100	1,400	50,000	495
San Carlos de Jonata	Sta. Ynez District.	26,634 31-100	100	1.500	100
Corral de Quati	Sta. Ynez District.	13,322 29-100	Stock in	c'd'd w'h	La Zaca.
La Zaca	Sta. Ynez District.	4,458 10-100	1,114	3,400	20
Canada de Los Pinos	Sta. Ynez District.	35,499	300	2,000	50
San Marcos	Sta. Ynez District.	35,573 10-100	Stock in	c'd'd w'h	Nojoqui.
Tequepis	Sta. Ynez District.	8,919	Stock in	c'd'd w'h	Nojoqui.
Los Prietos y Najalayegua	Sta. Ynez District.	48,728 67-100	200	None.	20
Las Lomas de la Purificacion	Sta. Ynez District.	13,341 39-100	1,000	3,000	10
Nojoqui	Sta. Ynez District.	13-284	1,315	5,950	88
Las Cruces (two leagues)	Sta. Ynez District.	8,888	100	5,900	50
		208,647 86-100	4,129	21,750	338
East ½ Nestra Senora del Refugio	Part of the Santa Barbara Valley and two of the Channel Islands.	13,264			
Canada del Corral	Part of the Santa Barbara Valley and two of the Channel Islands.	8,875			
Los Los Pueblos	Part of the Santa Barbara Valley and two of the Channel Islands.	15,534 76-100	2,500	22,200	250
Las Positas y La Calera	Part of the Santa Barbara Valley and two of the Channel Islands.	3,281 70-100			
Santa Rosa Island	Part of the Santa Barbara Valley and two of the Channel Islands.	62,696 49-100		29,000	100
Santa Cruz Island	Part of the Santa Barbara Valley and two of the Channel Islands.	52,760 33-100		27,000	100
		156,412 28-100	2,700	78,200	720

*Portions of the Guadalupe, Punta de la Laguna, Suey and the Cuyamas extend into San Luis Obispo county, but the FULL acreage is given here as represented by the U. S. patent.

CHAPTER III.

CLIMATE—REPORTS OF DR. NELSON AND DR. LOGAN—TESTIMONY OF OTHER PHYSICIANS AND VISITORS.

IT is impossible to say anything new about the climate of Santa Barbara. That it rains, sometimes, in the winter, that snow never falls, that even frost is a rarity,—all these things have been told a thousand times. Tables of temperature and rainfall, with other interesting matter, will be found at the end of this pamphlet. The figures tell the story as well as figures can. But it must be considered that not only these tables, but the printed praises of hosts of visitors, apply simply to the narrow strip of land noted as the valley of Santa Barbara proper, which lies south of the Santa Ynez mountains. Above Point Concepcion, the climate is colder and more moist. There the summer trade winds blow, from which this valley is protected by the mountainous islands in front and the mountain ranges behind it. The islands also intercept, to some extent, the rain-bearing winds of winter. The climate is still further modified by a warm ocean current flowing along the county's southern border. The interior valleys have an atmosphere remarkably dry and pure which in some diseases has been found of incalculable benefit. By the tables of rainfall for the past fifteen years, it will be seen that the amount is subject to great variation. It is in fact proverbial that no two years are alike. The present season (1883-4) has been the severest known for thirty years, 34.76 inches having fallen in this valley up to the first of May.

In order to "see ourselves as others see us," the remainder of this chapter will be devoted to the statements of the best authorities on climatology in its relation to disease; also the impressions of a few distinguished visitors.

Dr. Wolfred Nelson, C. M., M. D., Member of the College of Physicians and Surgeons, Quebec, Canada, at present U. S. Consul at Panama, South America, published an exhaustive article in the Planet, from which the following extracts are taken :

"This small but charming seaside city is probably the best known of all California resorts. Its fame, both for health and residence, is world-wide. * * * Owing to the bend of the Coast range of mountains to the north of the city and the nearer foot-hills, it is all but closed in by them. Many of the mountains, some distance back of the city, are 3,000 or 4,000 feet high. The valuable protection they afford the place will be appreciated at once by the medical mind. They deprive cold and damp winds of their moisture by filtering the air; hot breezes from the sandy plains, or the California desert of the olden writers, are likewise tempered by taking up a little moisture. They stand faithful sentinels to watch over the spot. * * * Owing to its geographical position, lying in a bend on the coast, the mountains to the north modify sea breezes and wind. Looking seaward, we see that its own coast is protected by the beautiful islands in the Santa Barbara channel. The large and mountainous islands of Santa Cruz and Santa Rosa, lie all but opposite, they are some miles out, and with others of the group form a protecting barrier that lessens the force of the vast rollers as they flow in from the grand old Pacific. * * * Santa Barbara has sprung into general fame within the last fifteen years, and now takes front rank, and deservedly so, as a health resort, and watering place. It has been selected by many experienced travelers as a home, both for residence and health; the scores upon scores of comfortable homes in the city, as well as the cosy suburban villas in the Montecito, surrounded by their sub-tropical trees, plants and flowers, say more for its healthfulness and delightful climate, than any mere verbal description can attempt. In that it is not expensive, it offers many inducements to those of easy or small means, who wish to settle down. The country itself and its environs have to be seen, and its climate experienced, to be awarded their due meed of praise. * * * A great deal has been written well and ably on the climatology of Southern California, but it will stand all the descriptions that have appeared hitherto, and still have a handsome balance in its favor for future writers, medical and lay. None know better than medical men than no one spot suits all classes of cases, or that no given climate is perfect. Patients must visit places like those now under consideration—and judge for themselves. It is to be presumed that all such leave home under medical advice. Their physicians will be able to say, generally, whether a stay on the sea-coast, or in the valleys or mountains of California, will be best suited to their cases. Sea air is often too stimulating for some pulmonary cases, while it agrees perfectly with others. Heat and a little moisture suit some cases very well, as we know, while the same agents exercise a pernicious influence on others.

Should mountain air be necessary, patients can visit the Ojai Valley, fifteen miles away from Santa Barbara, easily reached by the very comfortable line of stages of Mr. Rundell's line. There they can camp out or live at the hotel—2,000 feet above sea level. The Ojai Valley, however, is too important as a mountain sanitarium to be dismissed with this curt notice. It will be embodied in the next letter. There are hot sulphur springs six miles from Santa Barbara, very easy of access, in a cañon, with an hotel and bathing establishment. Board—including bathing—is some $14.00 a week. They are visited frequently by the residents as well as visitors. Apropos of board in Santa Barbara—it varies all the way from $7.50 to $18.00 per week, fires, when necessary on cold or chilly days, extra.

For centuries Mexico has had an unequalled reputation for all pulmonary and bronchial cases. California was originally a part of Mexico, the Southern part of the State on the coast joins the Mexican frontier—ten miles south of San Diego. It will thus be seen that California has inherited its good name for health. Many, many are living there now, as well as in various parts of the United States of America, who are indebted to its salubrious atmosphere for extended lives, and in other cases, a complete arrest of disease. Some far seeing people who have the hereditary tendency to phthisis, have sought such a climate with a view of aborting it in their children. That their forethought evinces the soundest judgment we must admit. There sick and well can almost live out of doors, constantly inhaling pure air. Under favorable circumstances, health is expected as a logical sequence. The days are enjoyable and the nights are cool; occasionally during the winter a fire may be necessary. Cool nights mean refreshing slumber.

It is far from the writer's intention to paint Santa Barbara as a perfect paradise on earth—such spots exists only in the mind of the poet. It, in common with all health resorts, has its drawbacks. Fogs at times, and occasional sand-storms, now and then very hot days (but not in the season), and once in a while cold, chilly days in winter, with a touch of frost. But even they have their advantages, as they remind one that there is no heaven on earth. Once over, the visitor or resident doubly appreciates his happy surroundings.

The "season" for visitors, sick and well, lasts from October to April or May, depending in a great measure on the severity, or early approach of winter in the Eastern States and Canada. Winter in Southern California extends from December to April, and means a small amount of rain, when good Dame Nature appears in her most becoming mantle of green, whose vivid hues are increased by wild flowers."

Dr. Thomas M. Logan, ex-President of the American Medical Association, and Secretary of the State Board of Health, made a statement in favor of Santa Barbara as a suitable place for a State Sanitarium. In his first official report, published in 1871, occurs the following: "The Secretary informed the Board that he had been occupied of late in visiting several localities in the southern part of the State, noted for salubrity, as San Rafael, Santa Cruz, Monterey, San Luis Obispo, Santa Barbara and other places. * * * While most of the localities named are possessed of climatic elements adapted to different stages and characters of pulmonary diseases, that of Santa Barbara appeared to present that happy combination of the tonic and sedative climate, which would seem to render it suitable to a greater variety of phthisical affections, and at the same time better adapted to the different stages of cachexia than any other place visited. For this reason he had pronounced it the most fitting point for a Sanitarium in California."

An article was prepared by Dr. Logan for the Scientific Press, giving a popular account of those leading features of the prevailing atmospheric constitution of the place, as modified by the physical conformation, and which go to make up what is here understood by the word "climate." In this article Dr. Logan says: "The very conformation and topography of this section, while it explains the cause, speaks to the intelligent reader of a climate that cannot be otherwise than even, mild and soft, and at the same time invigorating, with the moist but refreshing sea breezes which the thirsty land sucks in. In vain, heretofore, since my appointment to the responsible position of Health Officer to the State, have I sought for such a combination of sanitary qualities as are now presented. Here, in this mountain and island-locked valley, rising but a few feet from the blue waters of the grand old Pacific, all the prerequisites of health are to be found in measure so profuse that I would be accused of poetic extravagance were they duly portrayed. * * * As to the climate of Santa Barbara, it will be seen that, although lying in about the same latitude as Charleston, S. C., yet it is totally different, and that the isothermal line would be deflected towards St. Augustine, Florida."

Dr. Brinkerhoff, for more than twenty years a practicing physician of Santa Barbara, a man of a peculiarly philosophic and thoughtful temperament, offered some years before his death the following suggestions, which were incorporated in Dr. Logan's report: "Some ten miles from Santa Barbara, in the bed of the ocean, about one and a half miles from the shore, is an immense spring of petroleum, the product of which continually rises to the surface and floats upon it over an area of many miles. This mineral oil may be seen any day from the deck of the steamers plying between here and San Francisco, or from the high banks along the shore. Having read statements that during the past few years the authorities of Damascus and other plague-ridden cities of the East have resorted to the practice of introducing crude petroleum into the gutters of the streets to disinfect the air, and as a preventive of disease, which practice has been attempted with the most favorable results, I throw out the suggestion whether the prevailing sea breezes, passing over this wide expanse of petroleum, may not take up and bear along some subtle power which serves as a disinfecting agent, and which may account for the infrequency of some of the diseases referred to, and possibly for the superior healthfulness of the climate of Santa Barbara."

The freedom of Santa Barbara from epidemics, and its general healthfulness has been testified to, again and again. Dr. Brinkerhoff thus alludes to the facts of his own experience: "That the climate of Santa Barbara possesses elements of general healthfulness in an eminent degree, and perhaps, also, some latent peculiarities in its favor too subtle for ordinary observation, I may instance the following facts in this connection: During the eighteen years of my active practice here, I have never known a single case of scarlet fever or diphtheria. I have known of only three cases of dysentery, neither of which proved fatal; and of only three cases of membranous croup. The epidemics and diseases incident to childhood, which in other parts of the country sweep away thousands of children annually, are here comparatively unknown. Cases of fever and ague I have never known originating here, and persons coming here afflicted with it, rarely have more than two or three attacks, even without the use of anti-periodics. I have known instances of smallpox at three different times. In each of the first two instances, occurring several years apart, the disease was confined to a single case, and was contracted elsewhere. Neither of these cases proved fatal. In the year 1864, when the disease prevailed so extensively, and proved so fatal throughout the State, there were two cases contracted elsewhere and developed here, both of which proved fatal. Three other persons residing here contracted the disease from contagion at this time, all of whom recovered. Although no unusual precaution was taken to prevent the spread of the disease, it was confined to the cases mentioned. In the years 1869-70, when this disease, in its most virulent form, prevailed so generally throughout the State, not a single case occurred at Santa Barbara, although in daily communication with other points of the State by stage and steamer."

Dr. M. H. Biggs, for many years resident in Santa Barbara, and now a leading physician in Valparaiso, Chili, in his report to the State Medical Society, on the "Vital Statistics and Medical Topography of Santa Barbara," corroborates the testimony of Dr. Brinkerhoff in every respect. He says: "There are no malarious fevers. Persons who come here afflicted with fever and ague, rarely have more than two or three attacks. They soon become well, often without the use of anti-periodics. The climate seems sufficient to cure the malady. During a residence of over eighteen years, I have seen only one case of membranous croup, and heard of two others. There is no disease endemic to Santa Barbara—nothing but what can usually be referred, either directly or indirectly, to some indiscretion in eating or drinking, or unreasonable exposure."

Dr. E. N. Wood, the author of the first "Guide to Santa Barbara," published in 1872, says: "There are essentially two climates in California—the land and the sea climate. The latter derives its low and even temperature from the ocean, which along the coast stands at 52° to 54° through the year. The summers are hotter in the north. One might travel from Santa Barbara northward in summer for 300 miles, and find it hotter everywhere than here, or go southeast the same distance to Fort Yuma, one of the hottest places in the world. * * * In speaking of the "rainy season" we do not mean a season of constant rains, or anything like it. The term is employed only in contrast with the dry season and implies the possibility of rain rather than its actual occurrence. In this county, even in the seasons of most rain, by far the larger part of the winter is bright and clear weather. It is usually regarded as the most pleasant part of the year. It is spring rather than winter, and most of the rain falls at night. The grass starts as soon as the soil is wet, and at Christmas the land is covered with green. The climate is always kindly. We are not troubled by the hot, exhausting days of the Eastern summer, which have no respite at night. * * * We have no thunder storms, and the showers must needs be short and gentle that aggregate only twelve inches a year. From April to December there is no rain and one day is as another—bright, beautiful and life-giving. The gentle sea breeze is tonic and invigorating, and relieves the climate from enervation. The days combine the freshness of early spring in the Atlantic States with the softness and dreaminess of the Indian summer, and every day is a new delight. If one thinks of this continuing all the year with hardly twenty days' exception, he cannot doubt that Santa Barbara has a climate as nearly perfect as can be found."

Rev. J. W. Hough, D. D., writes as follows: "It is only just beginning to be known that Southern California has a climate whose dryness, uniformity, freedom from malaria, general tonic properties and fitness for outdoor life, alike in summer and winter, make it the sanitarium of the Western Continent for consumptives, and I might add, an admirable camping ground for the great army of over-worked, debilitated, nervous, sleepless men and women, whose ranks are constantly recruited by the devotees of business and fashion in Eastern cities. * * * In Southern California one may choose his climate. I have described the whole region as dry, mild and equable; but some points excel in dryness, and others in equability. One may live on the sea shore at Santa Barbara, or San Diego, or in the interior at Riverside; he may live on the plain at Los Angeles, or among the mountains back of San Bernardino. He may greatly vary his climate within a much smaller range. Here, in the town of Santa Barbara, in the Hot Springs cañon six miles distant, on the summit of the Santa Ynez mountains, twelve miles away, and at the newly discovered springs at the opposite foot of the range, twenty miles distant, are four quite different climates, the difference being effected by change in the altitude and distance from the sea. * * * "See Naples and die," say the proud Neapolitans. "Come to Santa Barbara and live," say the equally contented Barbareños."

Mr. J. C. Culbertson, in a letter to the New York Tribune, writes as follows: "During the winter there are twice the number of days on which an invalid or pleasure seeker can enjoy a walk in the sunshine that he would find at Pau, Nice, Mentone, Naples or Rome. This I know from years of personal experience. In these places the summer sunshine is almost unendurable, whilst in Santa Barbara it is the pleasantest season of the year. If the day is too hot, one has only to seek the shade and he will be fanned by a gentle breeze from the grand old Pacific. If too cool, he has only to step into the sunshine to be comfortable. However hot the days may be, the nights are always cool. Take the climate of Santa Barbara, with its rainy season and its dry season, its foggy days and its windy days, there are not more than twenty out of the 365 in which an invalid may not enjoy a walk in the sunshine some part of the day. And considering its winter, summer, spring and autumn together, I am sure there is not a spot on the civilized globe that will equal it in equability of climate and grand scenery and surroundings. But I would say to invalids: If you are far gone in consumption you need hardly expect to be cured anywhere; but by coming here and taking very great care of yourselves you may live a few months or years longer than elsewhere."

Ex-Governor Fenton, of New York, says of Santa Barbara: "It would hardly be poetic extravagance to say it is a fairy-like scene of land and sky, as inviting as this good earth can well be, with the fruits and flowers of a perpetual summer. * * * Its fascinations, especially as a health resort, it is not easy to overstate. * * * Speaking for myself, I confess that I overlook the drawbacks of a residence here, obviously there are some, for the climate, than which there can be no better. * * * Why not, when from my own observation in Florida, in Texas, the south of France and the world-renowned Italy, I can say, this climate at least excels them all!"

The Princess Louise of England, and her husband, the Marquis of Lorne, were free in their expressions of enjoyment of the beauty, rest and quiet of Santa Barbara. Their intention was but to remain a few days, but the time was extended to a fortnight, during which they drove or walked about town daily, the Princess sketching points of interest. They were especially pleased with the manners of the people, who allowed them to amuse themselves in their own way without intrusion.

Bishop John F. Hurst, D. D., LL. D., says: "You are sure to be conquered in California. If you do not surrender at one place, you will have to do it at another. In the North you can resist very well, even in the Yosemite; but you are in hopeless danger if you drift southward and sunward. * * * and that fair place overlooking the sea they call Santa Barbara, after her of wondrous beauty, who lived in the prison tower for her faith's sake, and died for it in savage old Bithynia."

Judge Freelon, of San Francisco, in a letter to a local journal says: "The idea that Mr. Nordhoff or anybody else has ever exaggerated the charms of Santa Barbara is preposterous! Many have tried to say something adequate on the subject and have failed ignominiously. Nobody yet has had the genius to portray Santa Barbara!"

Mr. J. R. G. Hassard, a staff correspondent of the New York Tribune, who spent several months in this country, says of Santa Barbara: "A placid little nook of ocean seems to have been provided by nature, where the sun shines almost always and the hurly-burly of traffic and speculation is hushed."

Testimony such as the foregoing can be gathered by the ton. It is a temptation to the writer to linger among the pleasant sayings and appreciative letters of such travelers. But a more practical duty lies before us, to enumerate the advantages which are calculated to win honest settlers, who may or may not feel any sympathy with the invalid class to whom this chapter is especially dedicated.

CHAPTER IV.

MEANS OF ACCESS—POPULATION—COUNTY PROPERTY AND DEBT—SCHOOL DISTRICTS AND THE SCHOOL CENSUS—MISCELLANEOUS STATISTICS.

THE means of access to Santa Barbara are as yet mainly limited to the approach by sea. The steamers of the Pacific Coast Steamship Co., Goodall, Perkins & Co., Agents, San Francisco, make regular trips—through steamers alternating with others that touch at several way ports. The through steamers, which touch only at Port Harford before reaching Santa Barbara, are the new and magnificent Santa Rosa, which has just been placed on the route, and the old favorite Orizaba. They sail alternately every five days from San Francisco, making the trip in about 30 hours. The Eureka and the Los Angeles sail alternately with the others, making a service of about fourteen steamers every month. But these latter boats carry freight mainly, to and from Santa Barbara, as they call at several ports each way, making the voyage somewhat longer for passengers. Freight steamers call at Carpinteria and Goleta, when sufficient freight is offered, as in the harvest season. The fare from San Francisco to Santa Barbara, and vice versa, is $10.00 in the cabin and $7.50 in the steerage. The steamers are all made as comfortable as possible, with experienced officers and obliging attendants. Besides the regular service, freight steamers, carrying combustibles, etc., sail every week for southern ports. The ticket agent of the company in Santa Barbara is J. A. Norcross; the local freight agent, John P. Stearns.

For those who dislike a voyage by sea, several alternatives are offered. A daily stage to Newhall via Ventura, carries the mail and connects with the Southern Pacific R. R., for San Francisco, or for Los Angeles, and from thence to Arizona and the East. There is also a daily accommodation stage to Ventura, a distance of thirty miles. To Newhall from Santa Barbara is about 85 miles. Many travelers going East prefer to take the steamer to Los Angeles, and from thence begin their long journey across the continent, by one of the southern transcontinental roads. Going north, a daily stage connects with the cars at Los Alamos, stages at San Luis, cars again at Soledad, to San Francisco. This is a particularly pleasant trip in the proper season, as it passes through some of the finest scenery in Southern California. The cars from Los Alamos also extend to the wharf at Port Harford, connecting with steamers up and down the coast.

The county is well supplied with wharves to facilitate communication by sea. Beside those on Santa Cruz and Santa Rosa islands, there is a wharf at Point Sal, the shipping point for Guadalupe and the Santa Maria valley; the Schute Landing on the Casmalia rancho; one on the Jesus Maria rancho and one at Purisima; one at La Espada and another at Gaviota, for the Santa Ynez valley, San Julian, etc.; at More's landing for La Patera and vicinity; Stearns' wharf at Santa Barbara and Smith's wharf at Carpinteria. All the landings below Point Concepcion are available in any weather, for although the so-called harbor is little more than an open roadstead, the islands protect it from the violence of heavy storms.

It cannot be denied that many have been deterred from settling here by reason of a lack of railways. For years the county has been vibrating between hopes of a railroad and fears of its effects. Periodical attempts have been made to induce some of the great corporations to build in this direction. In several instances we have almost heard the sound of the whistle piercing these mountain barriers—almost looked to see the smoky streamers of the locomotive descending to the sea. We have had railroad committees to make arrangements, and railroad engineers to report upon a practicable route—but no railroad, except the narrow gauge now resting on its laurels at Los Alamos. This will probably achieve the city in due course of time. But to satisfy in any degree the coming population, to whom railroads seem a matter of meat and drink, nothing will serve but a grand transcontinental, broad gauge, trunk line from the eastern seaboard, and a coast line from Alaska to Tierra del Fuego! Yet there are old residents who do not scruple to say that the bustle of coming and going trains, with their inevitable consequents, will take half the charms from this quiet valley.

In 1870, the population of Santa Barbara county, which then included Ventura, was given in the census reports at 7,987. In 1872 the county was divided, and in 1880 the census for Santa Barbara alone showed a population of 9,513. As a proof that the number is still increasing, reference is made to the school reports and the Great Register of the county. In 1879 the number of registered voters was 2,384; in 1882 it was 2,573; now, in 1884, it is estimated by the County Clerk, (no Great Register having been printed since 1882) to be at least 2,600 in round numbers, probably more rather than less. This shows a healthy increase. If the voting part of the population is, as commonly estimated, about one-fifth of the whole, the county may reasonably claim a population of nearly 13,000 souls.

According to the last report of the Superintendent of Schools for the year ending June 30, 1884, the number of children between five and seventeen years of age, amounts to 3,445; the number of school districts, 40. [This might be compared with the report of 1879, when there were but 30 school districts and 2,976 children of school age.] There are 13 Negro and 20 Indian children, counted with the above. In addition, the report notes 47 Mongolian children of school age, who are not allowed the benefit of the schools. The following are the names of the school districts, with the number of the school children in each: Agricola, 34; Artesia, 66; Ballards, 26; Bells, 84; Bear Creek, 27; Carpinteria, 113; Casmalia, 30; Cathedral Oaks, 66; Central, 99; College, 70; Dos Pueblos, 28; Guadalupe, 112; Honda, 16; Hope, 56; Jonata, 21; Las Cruces, 42; Los Alamos, 30; Lompoc, 195; Maple, 53; Miguelito, 23; Mission, 96; Montecito, 174; Najoqui, 19; Oak Vale, 38; Ocean, 59; Pine Grove, 41; Pleasant Valley, 40; Point Sal, 16; Purisima, 22; Rafaela, 78; Rincon, 80; Santa Barbara, 1,258; Santa Maria, 94; Santa Rita, 43; Suey, 44; Sisquoc, 23; Washington, 32; Laguna, 19; La Graciosa, 52; La Patera, 36.

The county has outstanding bonds to the amount of $46,500. Cash in the treasury, $34,318.75, to be devoted to the county's current expenses. The value of property belonging to the county, including a handsome brick Court House, is about $85,000. Last year the rate of taxation for the State and County was $1.694 on each $100; for the city, 85 cents. Legal interest is 7 per cent.; but the contract system prevails in this State, and money can be loaned upon good security at 10 or 12 per cent. per annum. This is a decided reduction from the rates of ten years ago, and it is probable that as the country becomes more settled and prosperous, interest upon money will still further decline.

CHAPTER V.

THE CITY OF SANTA BARBARA.

FRONTING the southeast, the city of Santa Barbara lies upon a sloping plateau, which, attaining within two miles a height of 300 feet, affords ample facilities for drainage. To the right, as the traveler lands upon Stearns' wharf, he sees the apparently low foothills, nestling at the base of a rugged mountain range that seems to bound the horizon; to the left, the mesa or table land, rising boldly from the plain and terminating abruptly in the foreground in the picturesque and often painted Punta del Castillo, sometimes incorrectly called Castle Rock. Between these, with the white towers of the Old Mission shining in the background, the beautiful city is outspread, covering a space of about two miles square, and laid out in generous blocks of 450x450 feet. It is a city of gardens. Every modest cottage is overrun with roses or covered by clambering vines, or encompassed with trees and tropic foliage, and blooming shrubs from every quarter of the globe. The garden of Dr. L. N. Dimmick, on De la Viña street, is especially noted for its collection of curious and interesting plants, translated from the most widely varying climates. Many of the streets are bordered with the graceful pepper tree, the elegant grevillea or the giant eucalyptus. Trees are subject to fashion, like everything else, and many varieties have had their day; but the beautiful pepper tree was the first love of Santa Barbara, and still holds its own against all comers. It is a tree beloved of artists and is sure of immortality upon canvass, at least.

Santa Barbara is an incorporated city of about 5,000 inhabitants. The present incumbents of the chief municipal offices are: Geo. W. Coffin, Mayor; J. R. Vance, Councilman for the First Ward; A. F. McPhail, Second Ward; J. N. Johnson, Third Ward; G. W. Leland, Fourth Ward, and I. K. Fisher for the Fifth Ward. The City Hall is a handsome brick building standing upon a small central plaza. In addition to rooms for the accommodation of the Council, City Clerk, etc., it provides quarters for the fire companies, two in number, a hook-and-ladder company, and "Washington No. 2" Engine Co., the latter company proud in the possession of a $4000 nickel-plated engine, which is mainly ornamental, as fires seldom occur. The principal buildings upon State street, the main business street of the city, are lit with gas, furnished by the Santa Barbara Gas Co. Water is freely distributed to all parts of the city through the mains of the Mission Water Co. This water is drawn from a clear mountain stream which descends through a narrow, picturesque cañon north of the city, called Mission Cañon, a resort in the season for picnic parties and excursionists who love Dame Nature in deshabille.

The public schools are of the first order; those in the Third and

Fifth Wards occupy fine roomy buildings, and all are supplied with excellent teachers. For all these reasons, private schools do not flourish. The only educational establishment of any size outside the public schools, is the St. Vincent orphan asylum, established in 1858, taught by the good Sisters, whose devotion to the cause of charity is understood and admired by Protestants as well as Catholics.

The city contains a fair proportion of church-going people, and several handsome church edifices. The largest congregation is that of the Catholic faith. Beside the Parish Church, on the corner of State and Figueroa streets, under the supervision of Rev. Father James Villa, services are held in the old Mission Church, by the Franciscan friars. Trinity Church (Episcopal) stands on Gutierrez street, Rev. John Bakewell, D. D., Rector. The Congregational Church, on Santa Barbara street, between Cota and Ortega, Rev. C. S. Vaile, pastor. The Presbyterian Church, on State street, is under the ministerial charge of Rev. M. L. E. Hill. Rev. A. W. Jackson preaches to a Unitarian congregation in Unity Chapel, on State street. Rev. W. A. Knighten is pastor of the Methodist Episcopal Church on De la Guerra street, and Rev. Dr. Nisbet supplies the Baptist Church on Micheltoreña street.

In respect to hotels, Santa Barbara claims a just precedence over all the other cities of Southern California. The Arlington, undoubtedly the finest, is really a magnificent house, thoroughly appreciated by tourists, under the management of W. N. Cowles. The Ellwood, lately opened to the public, is a roomy, comfortable, home-like hotel, managed by M. E. Hunt. The well known Morris House, J. F. Morris, proprietor, and the Occidental, C. H. Fiske, proprietor, are both handsome, well-furnished and popular hotels. The price of board at the different hotels varies from $6 to $20 per week, the latter price commanding an elegantly furnished apartment at the Arlington. Excellent meals can be obtained at Raffour's French Restaurant and the Central Restaurant. The proprietors of both can also furnish comfortable rooms for lodgers. There are several excellent chop-houses, coffee-houses, etc., where meals can be obtained for from 15 to 25 cents. These also deal in oysters, cooked to order, during the proper season. A number of private boarding and lodging houses, in different parts of the city, offer the very best accommodations to quiet guests, at varying prices.

Street cars run from the wharf to the Arlington. A popular mode of locomotion is the 'bus; several of these vehicles ply in all directions, or will even take parties upon short excursions. Fine hacks and carriages abound. Stylish livery turnouts and saddle horses can be obtained at reasonable prices at the following stables: Victoria Stables, J. N. Johnson; Arlington Stables, D. W. Thompson; Occident Stables, G. W. Leland; Champion Stables, A. F. Mc-

Phail; Fashion Stables, O. M. Covarrubias, and the Black Hawk Stables, D. W. Martin.

Wells, Fargo & Co.'s Express office is on State street; A. O. Perkins, agent. The Postoffice is a money order office, and P. J. Barber the reigning Postmaster. The Western Union Telegraph office is managed by Miss J. A. Norcross, and is opposite the Postoffice. There are two banks, the First National, W. W. Hollister, President, and A. L. Lincoln, Cashier; the Santa Barbara County National, W. M. Eddy, President, and E. S. Sheffield, Cashier.

Two newspapers provide the citizens with daily reports from the outside world. The INDEPENDENT is published every day, Sundays included, and a weekly edition Saturdays, by the Independent Publishing Co., George P. Tebbetts, Manager. The Press, daily, except Sundays, and weekly, by W. G. Kinsell.

Societies flourish. The following are the names of Lodges: Santa Barbara, F. and A. M.; Magnolia, F. and A. M.; Corinthian Chapter, R. A. M.; Marguerite Chapter, Order Eastern Star; Channel City Lodge, I. O. O. F.; Santa Barbara Lodge, I. O. O. F.; Seaside Lodge, I. O. G. T.; Santa Barbara Lodge, I. O. G. T.; Santa Barbara Lodge, Knights of Pythias; Santa Barbara Lodge, A. O. U. W.; Ocean View Council O. C. F.; Santa Barbara Council, A. L. H.; Starr King Post, G. A. R. The Santa Barbara Guard is commanded by Captain P. L. Moore. The W. C. T. U. have pleasant rooms on State street. Two Clubs, the Union and the Pioneer, have rooms handsomely fitted up, one in the First National Bank building and the other in the Odd Fellows' Building. There is a local circle of Chatauquans, a flourishing Natural History Society, Horticultural and Agricultural Societies. The two latter societies are very active and progressive, and their fairs are invariably occasions of special interest. Lobero's Theatre would do credit to a larger city than Santa Barbara, holding an audience of 1,200, with a fine roomy stage and provided with a fair amount of scenery. There are several associations of musicians, brass and string bands, orchestras, etc., whose members are much given to the practice of serenading. Almost any pleasant night may be heard the vibrant strains of wandering players making the soft air musical in spots.

A source of especial pride to the citizens of Santa Barbara is the Public Library. This was originally owned by the Odd Fellows; but a few ladies made heroic efforts in its behalf, whereby it was transferred to the city as a free gift. Friends of culture in and out of the city have made generous donations of coin and books, and 4,000 volumes now rest upon its shelves. The late Floral Carnival was one of many entertainments given for its benefit. It is managed by a Board of five Trustees, elected by the people. The principal periodicals and a variety of newspapers are received regularly. A movement is now on foot to provide permanent quarters, expressly arranged for the needs of the library. It is also proposed to extend its privileges, which have heretofore been limited to city tax-payers and visitors nominated by them. Since the people of the surrounding suburbs have worked hand-in-hand with the citizens, it is but just that they participate in its benefits.

Among the leading physicians of Santa Barbara are Dr. S. B. P. Knox, Dr. R. F. Winchester, Dr. Shaw, Dr. Bates, Dr. Crooks, Dr. Logan, Dr. Guild and a lady physician, Dr. Harriet G. Belcher. Among the legal fraternity, the following are the most prominent: Hon. D. P. Hatch, the present Superior Judge, Hon. Charles Fernald, late Mayor of the city, Messrs. W. C. Stratton, Paul R. Wright, E. B. Hall, R. B. Canfield, C. A. Storke, Chas. E. Huse, B. F. Thomas, J. J. Boyce, J. T. Richards, A. A. Oglesby, Thos. McNulta, R. M. Dillard, C. A. Thompson, R. H. Chittenden, J. H. Kincaid and A. T. Bates. Eight of these gentlemen have occupied the bench at times and are addressed as "Judge." The common Californian custom of bestowing the aforesaid title upon every member of the bar, does not obtain here. Three dentists suffice to keep the mouths of the people in order: Doctors L. G. Yates, D. B. Lee and H. W. Stauffer.

Following are the names of the principal business men of Santa Barbara: Real Estate and Insurance—Messrs. W. H. Woodbridge, Joseph J. Perkins, G. W. Coffin, A. O. Perkins and Fred. A. Moore. Photographers—A. M. Stringfield, C. W. Judd and W. J. Rea. Stationers—Alphonse Crane, H. A. C. McPhail and J. H. Summers. Lumber dealers—Gorham & Co., and Chas. Pierce. Furniture dealers and Undertakers—Knight & Blood, Robert Bell, F. F. Pierce, R. Forbush and A. H. Emigh. Carriage Manufacturers—Hunt & Son, A. C. Schuster and Joseph Bates. Druggists—H. J. Finger, A. M. Ruiz and B. Gutierrez. Jewelers—John Eaves, E. B. Chambers, Israel Miller, B. Guinand and P. Van der Linden. Dry Goods—N. P. Austin, W. E. Noble, H. F. Maguire, T. M. Breslauer and A. Garland. Dealers in Hardware—Edwards & Boeseke and Roeder & Ott. Boots and Shoes—Bell & Hunt, F. N. Emerson & Co., and T. Wharton. Clothing stores—Mortimer Cook, A. Fluehe, S. Hanauer, C. E. Hoffman. Grocers—John Walcott, P. N. Newell, Hunt & Metcalf, W. H. Myers, Smith & Johnson and J. H. Jacobs. Saddlers—S. Loomis, J. J. Eddleman. Milliners—Mrs. M. F. Hamer, Kenney & Hayward and Misses Ahern. Wholesale Butchers—T. K. Fisher and Sherman & Ealand. Candy Manufacturer—G. N. Johnson. Florists—John Spence and Joseph Sexton.

The cost of living here compares favorably with other places. If so fanciful a distinction might be made, one would say that the necessities of life can be bought cheaply and its luxuries dearly. A person of simple tastes can live in Santa Barbara perhaps at less cost than anywhere upon the continent. It is a delightful place for a small income; but for no income at all, the chances are not brilliant. A small house, unfurnished, can be rented for $12; a cabin for much less and a palace for much more. The prices of real estate are now at a reasonable level. A lot in the business portion of town would bring from $50 to $80 per front foot (on State street); a small lot in the residence portion is worth from $150 to $500; on the outskirts, good land can be bought for from $50 to $100 per acre, unimproved.

A steam planing mill, owned by Thomas Nixon, is kept running constantly. Mr. Nixon and Postmaster Barber are the principal architects residing in the city, and their work may be studied on many elaborate structures. The cannery, owned by the Dimmick, Sheffield & Knight Fruit Co., has already acquired a name abroad. Fruits, such as strawberries, apricots, nectarines, peaches, grapes, plums, pears, etc., are put up in the best manner. Tomatoes are also canned. This cannery, in the season, provides employment for from 80 to 125 persons. The city is well supplied with skilled mechanics. Laborers are also plentiful. If there be a lack of any kind of labor, it is in the line of domestic service. Chinese servants dominate many households. Female servants are very rare and highly prized.

The Hospital Farm, outside the city limits, was purchased several years ago at a cost of $9,000, and a great many necessary improvements have since been added. It is not intended as a refuge for paupers; but to shelter and care for the indigent sick. A poor house is an unknown quantity in this country. The necessities of the most poverty-stricken are supplied by a monthly sum, allowed by the Board of Supervisors upon proper representations.

Architecturally, Santa Barbara may be called of a mixed composite order. Pretty bay-windowed cottages are common. More pertentious dwellings are not rare. Of these the most palatial is the magnificent new residence of Thomas B. Dibblee, built upon the Punta del Castillo, a fine point of vantage. The grounds are laid out in the most generous, yet artistic fashion, and the building itself is one of the richest in Southern California. The residence of Don Gaspar Oreña, near the Mission, is also noticed by strangers for the beauty of its architectural design, and others, the property of John Edwards, J. W. Calkins, J. W. Cooper, Judge Fernald and Captain W. E. Greenwell, would be considered handsome buildings anywhere.

Of society, as the word is usually defined, but little can be said. When it is considered that a moiety have strayed hither from all points of the compass; meeting here the pastoral princes of a post-platal period or the children of a still earlier age—*hijos del pais*—could any homogeneity be expected? Social divisions are mainly made upon religious lines, though not to so great a degree as formerly. There is some influence—doubtless climatic—which renders the manners of the people at large more gentle and gracious than is apt to be the case elsewhere. This peculiarity was, strangely enough, first noted in print by a traveling humorist, "Derrick Dodd" of the San Francisco Post. And the inevitable admiration for the country which infects society to a remarkable degree, gives rise to a certain clannish temper, exasperating to tourists of the *nil admirari* type.

CHAPTER VI.

EL MONTECITO.

THE word "monte" means a thicket, and land covered with brush is in this part of the country known as "monte land." It is sure to be fertile when cleared. Montecito therefore means only a diminutive monte, and whoever may have driven through to the old Dinsmore place twelve years ago will recognize the aptness of the name. Strangers usually jump to the conclusion that the word means "little mountain," from the similarity of sound. El Montecito is simply a suburb of Santa Barbara. It is in a charming valley about four miles eastward from the city, and contains many handsome residences. The valley opens to the southwest, on the sea. The farms are mostly small, but in the highest state of cultivation. A large number of places rank as country seats rather than as farms. Among the finest places, occupied by capitalists or professional gentlemen, is the beautiful residence known as the Eddy place, now owned by Mr. Larminie, an English gentleman. Fronting this, is Judge Hall's new country seat; not far from his old place, now owned by J. M. Forbes, the well known railroad man. Mr. Forbes has bought largely in Montecito, and possesses a handsome estate there which is being rapidly developed. Other elegant country seats are those of Col. Hayne, I. R. Baxley, Dr. E. W. Crooks, J. M. Parks, Mrs. Dinsmore, J. Doulton, H. Stoddard, R. Kinton Stevens, Wm. Field and others. Montecito is especially adapted to horticulture. Many tropical fruits come to perfection. Among others, the bananas, planted by the late Col. Dinsmore, still ripen freely in the warm air. The place is now owned by Messrs. Johnston and Goodrich. The series of experiments in horticulture made for years in this valley by Col. Bond, who still retains his magnificent place, and Col. Dinsmore, have been invaluable to those who came later. The flourishing orchards of these two places, with those of J. M. Hunter, O. A. Stafford, H. L. Williams, H. C. Thompson, Col. Hayne, C. Aug. Thompson and others, show what delicious fruits the kindly soil can produce.

The Montecito Nursery, under the management of J. D. Reeves, is an interesting place to visit. It lies in a sheltered spot, where the most delicate plants can be coaxed into life. The whole valley is so situated that it is usually a few degrees warmer than in the neighboring valleys. Strawberries are particularly productive. The only place in the valley that could be denominated a large farm is the Swift place, which occupies a commanding position; the only rancho the Jacques place, now owned by H. L. Williams, whose acres are spread out over Ortega Hill, and support quite a number of hogs and other live stock. Within the last year so much property has changed hands and so many buildings have been erected that the Montecito would scarcely be recognized by an old inhabitant. The old Matanza property, once covered with fine oak woods, has been cleared and planted with fruit trees in profuse number. As an industry of Montecito, perhaps bee-keeping is quite as important as any. Every cañon has its rows of bee hives, and upon almost every miniature farm the bees swarm in myriads. There has been at times some unfriendly feeling between the bee men and the fruit men, the former's protegés being accused of eating some fruits before they were fully ripened, and even attending to the grapes while in blossom.

The former attraction of Montecito, the big grave-vine, was taken up and carried to the Centennial in sections. But the "daughter" vine bids fair to eclipse its parent. The "big grave-vine" property is owned by Dr. Doremus. Every one, who would be likely to read about Santa Barbara at all, must be fully acquainted with the story of the vine—how it was planted by a lovely Señorita, to whom it had been given by her lover, as a whip, and how it grew and bore tons of delicious grapes—so we spare further recital of the well-worn tale. The trellis supporting the vine made a shady saloon beneath, which for years was used by the native population as a dancing-floor. In Dr. Wood's Guide, alluded to previously, the following poetical notes occur: "Under this tropical trellis, on the hard beaten earth, many a rising moon has thrown its level beams upon Señor and Señorita dancing to the sad guitar, or upon the easy swing and wild abandon of the Spanish fandango. Here is the romance of the south. Here should the poet lounge and smoke in the starlight without, watching the dreamy convolutions of the waltz, listening to the soft rhythm of the Spanish tongue as voices float above sighing of the music. Here, too, have come the murderer, and the outlaw, stopping for pleasure in their flight from death. And here the officers of justice overtaking the fugitive, a desperate shot and its answer have interrupted the dance for a moment and put the desperado beyond reach of judge and jury."

Except the school buildings in the central part of Montecito, and a small Catholic church, where services are held occasionally, there are no public buildings of any kind. For church-going, marketing and social amusement, the residents depend upon Santa Barbara. There are quite a number of native Californian families in the valley, whose ancestors, as far as they know, were born on the same spot. They are generally a simple and kindly people who care little for things outside the circle of their native mountains. As a curiosity of the last census, it may be stated that there are 68 persons claiming the name of Romero. A description of the Montecito Hot Springs will be given elsewhere.

CHAPTER VII.

CARPINTERIA.

THE valley of Carpinteria was a part of the pueblo lands of Santa Barbara. In early days it was the custom of the prefect (an officer equivalent to the Mayor of to-day) to portion out the tillable lands among the people, who raised upon it their summer crops and then retreated to town for the winter. No titles to the soil ever passed before the coming of the Americans. The name was derived from a traditional carpenter, whose place of business was called in the Spanish tongue, el Carpinteria (with the accent on the penultimate). The thickly settled central portion of the valley is twelve miles east of Santa Barbara. The silver beach curves gently from the point dividing Montecito and Carpinteria, to the bold and rocky Rincon, giving the whole valley a southerly exposure. From the beach, it seems as if the land was enclosed, from point to point, by a deep semi-circle of purple mountains. In this well protected corner lies about ten square miles of deep, fertile, mostly alluvial soil. These rich bottom lands are the natural home of the Lima bean, Carpinteria's chief export. Fruits and nuts are produced in quantities and attaining sometimes an abnormal size. One of the first American settlers tells us that in early days it was not thought that this valley would ever be valuable for horticulture, owing to the utter absence of large streams for irrigation. But in time it became thoroughly apparent that the natural moisture might be retained in the deep loamy soil, by proper cultivation, making any artificial irrigation superfluous. Now, in the heat of summer, may be seen green fields of corn or beans, stretching away for miles, sometimes only divided from each other by a narrow roadway; fields on which no rain has ever fallen, and the nearest water is sixty feet under ground. A full account of the Lima bean industry will be given in another chapter.

Low foothills, at the base of the mountains, are sometimes cultivated to their very summits. On mesas and rolling lands, wheat and barley produce heavily and great crops of hay are harvested. Here and there, along the face of the apparently solid range, open mysterious cañons, watered by trickling brooks from distant springs. The enthusiast who attempts to follow one of these to its source, after tramping for miles along the tiny stream will be apt to give up at last in despair. He will admire the gay luxuriance of wild growths; the sycamores with their silver-grey trunks and roots sprawling in the gurgling water. If it is the right season, the delicious wild blackberry will tempt him always a little further. And if he is a botanist, he will be fully recompensed for his fatigue in the presence of the most delicate and fairy-like of ferns. Looking from the sea-beach toward the mountains, one would say that these endless cañons were not possible, and indeed he who has never traversed their length will be unable to believe the half that might be told.

Carpinteria is divided generally into small farms; a few acres of this astonishing soil being enough for the support of a family. Prices are high by comparison—but by actual worth, low, the best alluvial lands bringing from $100 to $200 per acre. Mesa land, suitable for grain, is worth from $5 to $30 per acre. All the best of the cañons, which are generally government land, have been taken up, but some of these claims can be bought very reasonably. The chief product of these cañon farms is honey, bees seeming to take kindly to the wild flower food in the vicinity.

The Carpinteria wharf, a good solid structure, is owned by the Smith Bros., whose house, lumber-yard and warehouses serve to give the place a busy air. The produce of the valley is here shipped to its market. A few miles beyond the wharf is the brick store of Mr. Thurmond, which also serves as the postoffice, and is surrounded by a cluster of buildings, among them a restaurant and blacksmith's shop, with several dwelling houses, making quite a village-like appearance. The hamlet also contains in a somewhat scattered way, two churches, Presbyterian and Baptist, two school-houses, a hall for meetings, etc. There are lodges of Good Templars and of the Knights of Pythias. The pride of the community, however, is a flourishing brass band, known far and wide as the "Lima bean band."

Near the center of the valley, the highly-cultivated lands and handsome residence of Col. Russel Heath is one of the most interesting points in the whole valley of Santa Barbara. Here were the first successes of the splendid English walnut tree, Col. Heath having been for several years noted as the largest walnut-grower in the United States. He has also extensive orchards of other fruits, a large quantity of peaches, apricots, etc., being dried every season upon the place. Other valuable orchards are those of O. N. Cadwell (called Pomona's Retreat), E. H. Pierce, J. A. Blood, Sr., E. J. Knapp, Mrs. M. A. Ashley, F. H. Knight, T. L. Knap, Seth Olmsted and others. The walnut grows very slowly, not beginning to bear until ten years old.

After that, every year increases the nut crop, and as a provision for his posterity, no man can do better than to leave each of his children ten acres, or even five of the English walnut. Citrus fruits do fairly well in some places. Prunes have been found by Mr. E. J. Knapp, the principal person who has tried them, to be an excellent paying crop. Apricots have been a favorite fruit, bringing good prices and selling well in a dried form. Peaches, nectarines, figs, olives, pears, apples, quinces and peanuts do remarkably well. Besides the Lima bean, the ground produces tremendous crops of other varieties of beans, corn, squashes, etc. Flax thrives, but is objected to by some farmers on account of the rapid exhaustion of the soil on which it grows. It is raised for the seed only; the fibre is not utilized. The farms of Messrs. T. A. Cravens, H. Lewis, the Franklins, the Thurmonds, the Ogans, Messrs. Thomas Pye, Henry Fish, L. B. Hogue, P. C. Higgins and others, who are generally largely interested in Limas, are perhaps representative places.

The perfume farm of H. H. Hall has been an object of much intelligent curiosity. Roses, violets, jasmines, tube roses and orange flowers are here cultivated for the express purpose of distillation for perfumery. The plantation is now doing well, and in due course of time the object will be accomplished, and perfumery become an export of the country. The beautiful grounds of H. C. Ford, the artist, contain a collection of plants similar to that of Dr. Dimmick of Santa Barbara, in respect that they have been brought from all parts of the world to ornament his garden.

One of the chief delights of life in the Carpinteria is known as "going a clamming." Sometimes, when the tide is out, the clams may be gathered by the bushel, with no aid except case-knives, fingers and toes. At other times they lie deeper, and parties come down to the shore with a team of horses and a plow; a furrow is slowly opened, and one follows with a basket to pick up the astonished clam. Without fear of contradiction, we assert that these Carpinteria clams, purified by a wave or two and roasted on a fire of drift wood, on the beach, and taken immediately, are good!

In a thickly settled valley, like that of the Carpinteria, which contains a population of over 700 persons, mostly farmers, it is impossible to do more than glance at a few—not necessarily the most notable—just to give an idea to the stranger who reads of the earth's marvellous capabilities. From the splendid acreage of the Bailard ranch to the smallest door-yard, everything speaks of thrift and energy, with their natural consequence of prosperity. And no one can wonder that the inhabitants of the rich valley of the Carpinteria are the most contented race the world can show.

CHAPTER VIII.

GOLETA AND LA PATERA.

GOLETA is situated about eight miles west of Santa Barbara, and consists of a store, postoffice and public hall, blacksmith shop, wagon shop, boot shop, school house, M. E. Church, a Baptist Church just finished, an eating house, and about half a dozen dwelling houses. It has two daily coaches carrying the U. S. mail, and can boast of a public watering trough fed by a wind mill and tank. There are two principal roads, the one running through the town from east to west and being the county road from Santa Barbara to Lompoc by the way of Gaviota, and the other called the stage road, running north across the mountain to the town of Santa Ynez. The entire surrounding region for a distance of several miles is commonly known as La Patera, a Spanish word which means "the duck pond." The shipping is done from More's wharf, situated about one mile south of the store. It is a commodious, well built structure, fully equal to the requirements of the farmers. D. M. Culver is the wharfinger. In the immediate vicinity of the wharf is an extensive deposit of asphaltum, of excellent quality, which is shipped in large quantities to San Francisco. There are also several clam beds not far from the wharf, which are chiefly valuable to pleasure-seekers. The election precinct, of which Goleta is the polling place, comprises three school districts, Rafaela, Patera and Cathedral Oaks. At the last general election there were one hundred and twenty-five votes cast. The entire number of children between the ages of five and seventeen, according to the last census reports, aggregates one hundred and eighty. The total population probably amounts to eight hundred.

The social advantages are good. The schools are ably managed, and largely attended. The churches and Sunday Schools are well sustained. A flourishing lodge of Good Templars exists; and only one whisky saloon has been able to withstand the better influences which prevail. The soil is deep and fertile. The bottom lands sometimes produce four tons of hay per acre. One ton of beans per acre is not an uncommon yield. Squashes of enormous size, are a specialty. They frequently exceed two hundred pounds in weight. A few years ago, one produced by Philander Kellogg reached such gigantic proportions, that when bisected, the cavity was found to be sufficient to allow the halves to be placed together enclosing his 18-year-old daughter. Hence the story has gone forth that Goleta squashes sometimes contain 18-year-old girls. This has naturally led to numerous requests from parties at a distance (presumably bachelors) for some seed of that remarkable variety. Large orchards of apricots, peaches and English walnuts have recently been planted, the perfect adaptability of soil and climate having already been demonstrated on a smaller scale.

A very important industry is the pampas plume culture, introduced here by Joseph Sexton. Almost incredible profits have been realized by those engaged in it. The principal producers are Joseph Sexton, Frank E. Kellogg and Chas. Hails.

For dairying purposes this section possesses unusual advantages. Owing to the richness and moistness of the soil, aided by the summer fogs, an abundance of green feed can be had at all seasons of the year without irrigation. There is a good supply of timber, principally live oak. Goleta is one of the chief sources on which Santa Barbara depends for fuel. The cost of stove-wood on the ground, is usually about four and one-half dollars per cord. At present there is for sale near Goleta, about eight hundred acres of land, ranging in price from one hundred and fifty, all the way down to five dollars per acre, depending upon quality of soil and location.

The scenery is very desirable. On the north are the rugged Santa Ynez mountains and on the south a low range of hills called the Mesa, which with the exception of here and there a gap, hides the ocean from view. There are several mountain streams, some of the cañons of which contain charming little falls. To obtain a most magnificent view, one needs only to take the stage road and go to the summit of the mountain. Then with Santa Barbara and its surroundings to the east, the rolling hills which stretch away toward Gaviota to the west, the loveliest of valleys nestled at his feet, the glimmering waters of the Pacific beyond, with the dim blue islands in the distance, and behind him a wilderness of mountain peaks, he commands a scene of surpassing beauty and grandeur. It is a scene which that extensive traveler, the late Secretary W. H. Seward, is said to have pronounced the most magnificent he ever beheld.

Among the most highly improved large ranchos are those of Ellwood Cooper, W. W. Hollister, S. P. Stow and John D. Patterson. There are many smaller farms that are also well improved. Among them are those of Joseph Sexton, B. F. Pettis, J. O. Williams, Henry Hill, I. E. Martin, George Edwards, A. C. Scull, F. E. Kellogg, Sr., A. E. Hollister, B. F. Woods and Charles Hails.

CHAPTER IX.

COAST RANCHOS—POINT CONCEPCION.

BEYOND the fine rural settlement of La Patera, the broad western avenue is continued to the rancho lands of Col. W. W. Hollister. His principal entrance gate is about three miles from Goleta and the ranch house and buildings occupy a pleasant eminence, commanding a fine view of the surrounding country. The family home, however, is reached through a narrow valley which at the end of two miles terminates apparently in a broad level opening, where the house stands, embowered in bloom and verdure, with its splendid orange orchard in front, and a fine profusion of other fruits clustered about it. This lovely retreat is named Glen Annie, in honor of the Colonel's charming wife. The entire rancho covers 3,600 acres, which allows for gardening on a generous scale. It embraces valley, mesa and grazing lands, and fronts the avenue for a mile and a half; running back to the mountains, through Glen Annie, about 3½ miles. For years Col. Hollister has spent time and money on experiments in agriculture, to prove for the general benefit, of what the country is capable. Twelve years ago, a correspondent of the New York World was inspired to utter the following prophecy, which has been fulfilled, more or less: "As I left this inviting spot I ventured to predict that ten years hence the Santa Barbara valley for sixty miles along the bay will, by reason of its unsurpassed climate and fertile soil, have become a very garden, in which this beautiful farm will stand prominent as the gem of Southern California, an example of what beneficent Nature and the skilled hand of one of her worshippers can do toward creating an earthly paradise."

"Ellwood," the celebrated olive-oil rancho, the property of Mr. Ellwood Cooper, joins Col. Hollister's on the west, and embraces about 2,000 acres, with a frontage on the avenue of three-fourths of a mile. This estate is noted also for the long lines of giant eucalypti, which mark the boundaries and division fences. It is said that Mr. Cooper planted 150,000 of these trees with his own hands. He has written a valuable book on the eucalyptus tree, recognized as a standard work. But his enthusiasm is for the olive, and he is the author of an exhaustive treatise on olive culture and olive oil, which said oil is produced on this rancho of most marvellous purity. He is also the President of the State Horticultural Society, and is particularly noted for the tireless energy with which he battles against pestiferous insect tribes, the special bane of horticulture in California. "Ellwood" is a favorite drive from Santa Barbara, and no tourist has done the country until the Hollister and Cooper ranchos have been visited. The latter must not be understood as covered with olive and eucalyptus trees. There are carefully tended orchards of citrus fruits, almonds, walnuts, and a great variety of other fruits which thrive throughout the valley.

Adjoining Hollister's on the east, back from the avenue, are the highly developed farms of W. W. Stow, of San Francisco, and his son, S. P. Stow. They own about 1,200 acres of farming, fruit and grazing lands. Mr. Joseph J. Perkins, in his "Business Man's Estimate of Santa Barbara County," thus notes an indubitable truth regarding the three estates just mentioned. He says: "What has been accomplished by Messrs. Cooper, Hollister and Stow in these three cañons can be repeated, with enterprise and means, in any or all of the cañons along the coast line of the Santa Barbara valley."

For some distance, the coast is a mere succession of cañons or rather cañadas, since they are often of a generous width. Las Armas, branching out from the "Ellwood" rancho, is a delightful nook.

The next cañon, the Tecolote, is one of the most naturally beautiful of all. It originally was a part of the Dos Pueblos, and is now the property of the Sturges brothers.

Eagle cañon, containing about 1,200 acres, has recently been purchased by Mr. Isador Dreyfus, for $30,000.

The rancho known as "Los Dos Pueblos," in remembrance of the time when two Indian cities were situated upon it, was patented to Nicholas A. Den for over 15,000 acres. But the heirs of Mr. Den now hold but 8,285 acres of the original grant. Upon this portion, there is a fine fruit orchard, but the land is mainly devoted to stock-raising. Near Dos Pueblos, to the west, is the dairy farm of A. W. Buell, who has occupied the place for several years. The next rancho, called Cañada del Corral, is owned by Señor Don Bruno Orella, and is chiefly devoted to sheep.

The Tajiguas rancho of about 2200 acres, lately owned by the Young brothers, has recently been purchased by Lawrence W. More, the price being $22,000. Upon this rancho are a number of bearing olive, orange, lemon and lime trees, of the choicest varieties.

The rancho of Our Lady of Refuge, Nuestra Señora del Refugio, owned by Col. Hollister and others, covers most of the remainder of the littoral belt under the protection of St. Barbara. In the midst of this "refuge," is the Gaviota landing, with a wharf and Postoffice. The road, which has followed the beach for miles, now turns sharply to the north and in the course of a mile or two reaches the Gaviota Pass; a natural chasm about sixty feet in width, with its walls of solid rock rising almost perpendicularly. Here the road crosses the Santa Ynez range, and descends to the valley of Santa Ynez

A few miles beyond Gaviota is Point Concepcion. The Coast Survey description calls it "a high promontory, stretching boldly into the ocean and terminating abruptly." It is more than two hundred feet above the ocean, and is conspicuous from land or sea in every direction, for many miles. The view from there is widely extended and magnificent. Here the government has erected a beacon light to warn the mariner far out at sea. The light itself is two hundred and fifty feet above the water and can be plainly seen from the heights behind Santa Barbara, more than forty miles away in a direct line. From this direction the point appears like an island. The illuminating apparatus is of the first order of the system of Fresnel, and exhibits a revolving white light, showing a flash every half minute. The lighthouse is a brick building, plastered, with a low tower rising from the center, also of brick, covered with white plaster. The government requires a daily record of observations to be kept, and the keeper and his family are also expected to extend the hospitalities of the house to visitors, besides keeping the light and lighthouse in order, and holding themselves always prepared to relieve any ship-wrecked crews who might be cast upon the point. The fog-whistle is located on the edge of the sea wall, directly below the lighthouse, and the operating machinery is so situated that the waves dash over it. During thick weather the whistle is heard once in every 52 seconds, night and day. A fog bell is also used, which weighs 3136 pounds, and is sounded every 13⅓ seconds in foggy weather.

This is the end of the Santa Barbara Valley!

CHAPTER X.

SANTA YNEZ.

THE upper valley of the Santa Ynez comprises 223,185 acres, according to the table on page 12, and it is estimated that not less than 50,000 acres are adapted to agriculture and horticulture. Since the table alluded to was compiled, the valley has been put under cultivation to a large extent. It is estimated that the grain shipped from Santa Ynez this year will amount to 30,000 tons, of which the greater portion is wheat. It is, in fact, a wheat-growing district, and is destined at some future day to furnish homes for thousands of happy families. The river first makes its debut in the county of Santa Barbara, on the wild and mountainous rancho called Los Prietos y Najalayegua. This is a region of cinnabar, and quicksilver mines were once opened, but have been allowed to relapse into ruins. It is now chiefly interesting to sportsmen, although many of its small valleys are susceptible of cultivation. The San Marcos rancho belongs to the Pierce brothers of San Francisco, and contains a large proportion of very excellent land; but it is at present only used for stock-raising. On the opposite side of the river, the Tequepis rancho is under the same ownership. Next to the San Marcos, the Cañada de los Pinos, known as the College ranch, is the property of the Catholic Church. The "College" which is responsible for the sobriquet, a school for boys, was kept up for several years, taught by priests of the Catholic Church. The rancho is managed under the direction of the Catholic Bishop of Southern California. A large portion of it is leased to various tenants, and brought under cultivation. Some of the choicest land, purchased by Mr. Cornelius Murphy, has been built upon and otherwise improved. It is mainly devoted to wheat and the raising of fine stock.

About two miles from Mr. Murphy's place is the little town of Santa Ynez; a nucleus, of which great things are prophesied. It is two or three miles from the river, and perhaps thirty-five from Santa Barbara, by way of the toll road over the San Marcos Pass. It contains a postoffice, express office, blacksmith shop, hotel and two stores, one belonging to Mr. H. Watkins, and the other a branch of the Goldtree firm of San Luis Obispo.

At a short distance from this little town the Hayne Bros. are devoting themselves especially to olive culture. Several thousand trees have been transplanted from Montecito and set out upon their ranch this year, and are said to be doing admirably, as well as those planted in other years. At present they are raising also considerable wheat.

Little is left of what was once the Mision Santa Ynez, in this vicinity. But in the least ruinous of the buildings a stock of general merchandise is displayed for the benefit of the neighboring farmers. About three miles from the Mission and three from Santa Ynez, the three places forming the three corners of an ideal triangle, is the rudimentary settlement called Ballard's. Here the main roads meet; one connecting the valley with the county seat by way of the Gaviota Pass, and one passing over the range at the San Marcos Pass, which, at its summit, is just fifteen miles distant from Santa Barbara. The land in the neighborhood of Ballard's is of excellent quality. Several thousand fruit trees have been set out here within the last two years. There is a postoffice, store, blacksmith's shop and school house at Ballard's.

Opposite the College Rancho, between the Santa Ynez river and mountains, is the Lomas de la Purificacion rancho, owned by the heirs of Captain Thomas W. Moore. It is a fine property, devoted mainly to stock raising. Following the river, on the same bank, is another rancho belonging to the Pierce brothers, called the Nojoqui.

The San Carlos de Jonata is the only remaining rancho of great extent in this district. It is composed partly of fine bottom land, partly of rolling and hill lands; much of it is well adapted to grain. It is well watered and for dairying purposes has no superior. The Childs postoffice, on this rancho, is about five miles from Santa Ynez. There is also a store and a school house in the same vicinity.

One of the finest characteristics of the upper Santa Ynez Valley, as a whole, is the abundance of pure clear water. The river itself and numerous little creeks that empty into it on either side give the land an exceptional value. Its horticultural capabilities are just beginning to be understood. Apples, pears, peaches, quinces and the small fruits do well. Prices have advanced somewhat in the last year or two. Fine land can no longer be bought for a dollar an acre. In fact, it now varies from $5 to $40 per acre. This is a wide variation; but the quality of the land varies widely, from good grazing to the best wheat and fruit lands. Among the causes which have contributed to the advance, may be numbered the narrow gauge railroad at their doors, the wakening of owners to the true value of their property, and lastly, the comparatively small amount in the market for sale.

CHAPTER XI.

THE LOMPOC TEMPERANCE COLONY.

THIS, one of the most famous temperance colonies in the United States, was founded in the Fall of 1874, in the most westerly portion of Santa Barbara county. The history of this colony is briefly told. In the Winter of 1869, while on a business trip to Santa Barbara county, Mr. W. W. Broughton, now residing at Lompoc, conceived the idea of organizing a temperance colony in Santa Barbara county if lands suitable could be procured. About the middle of December, Mr. Broughton met with George H. Long, then Superintendent of the San Julian, Lompoc and other ranchos devoted to sheep and owned by the Messrs. Hollister and Dibblees. Mr. Long was favorably impressed with the idea and enthusiastically recommended the Lompoc valley as the place of all places for such a community to locate, and on the day following the interview Mr. Broughton visited the property. Finding it all that Mr. Long recommended it to be, and more, Mr. Broughton immediately returned to Santa Cruz, his home at that time, and enlisted Judge E. H. Heacock in the enterprise, who immediately opened a correspondence with Col. W. W. Hollister, relative to purchasing the Lompoc rancho for colony purposes. The Colonel responded that the rancho had not yet been patented, but as soon as the title could be perfected it would be for sale. Something like one year passed before anything further was done; when the Colonel wrote the Judge that the title was perfected and the property, comprising about 47,000 acres, was for sale at $300,000. Terms were immediately agreed upon and a stock company called the "Lompoc Valley Land Company" organized, with a capital stock of $300,000. The stock was divided into one hundred shares of $3,000 each and soon sufficient was subscribed to warrant a forward movement, and a preliminary meeting was called to learn the object and aim of each subscriber. At that meeting it was developed that the largest stock holders designed going into sheep and cattle raising, which would leave the farmer to fence or to be continually annoyed with the herds about him. This meeting was anything but harmonious and resulted in the withdrawal of those intending to embark in general farming, and the enterprise was abandoned. Up to this time the agriculturist had no protection, except by constructing expensive fences, and this was out of the question in the greater part of the State, and especially so where stock growing was the chief industry.

Col. Hollister had for years been advocating a trespass law, that would compel the stock growers to fence or herd and allow the farmer some chance to build up a home unmolested. As soon as the Legislature went into session, the Colonel procured the passage of a trespass law that abolished the necessity of expensive fencing to protect crops. And from that law dates the new era of progress and prosperity of Southern California, where the law was made applicable.

Immediately after the passage of this law, negotiations were again opened for the purchase of the ranch and the terms agreed upon were $500,000; the trespass law enhanced the value of real estate two-fifths and it found ready purchasers. Suffice it to say, that in a few weeks the stock was all taken and the company fully organized and surveyors set at work subdividing the lands. The grand sales to stockholders—for no others were permitted to bid—commenced on November 9th, 1874, and continued for four days; when it was found that less than one-third of the property had been sold for $750,000.

Immediately after the sales the colonists commenced moving in and soon two hundred families were domiciled within the colony, and the great majority tenacious of the principles on which the colony was founded. But this basic principle was destined to be infracted and early in the history of the colony we find some fifty or more ladies one afternoon entering the colony drug store and summarily suppressing the traffic that had been illicitly conducted. Later on, a saloon was blown to atoms, and lastly, one was torn down which resulted in the prosecution of one of the leading citizens at Santa Barbara, when a jury of twelve men, within one hour after the case was submitted, brought in a verdict of not guilty. Thus, it is to be hoped, has ended the business of transgressing the principles of the colony, which now numbers between four and five hundred families, constituting one of the most intelligent and moral communities in the State.

There are nine schools within the colony, with a magnificent school edifice in the town proper. The religious sentiment is represented by most all denominations, the chiefest being the North and South Methodists who have each fine commodious churches. The town has all the needed appliances for progress and quick interchange—a daily mail, express and telegraph—with several mercantile houses of all kinds, two physicians, one drug store, two hotels, one jewelry store, one harness shop, two livery stables, and two blacksmith shops, with two good wharves over which freight is received and shipped. The town is located 16 miles from Los Alamos, 30 from Santa Maria, 28 from Guadalupe, 25 from Santa Ynez, and 60 from Santa Barbara. The present means of reaching the colony is by steamer to Gaviota, thence by stage to Lompoc, or by steamer to Port Harford, thence by rail and stage to Lompoc. There is a good prospect of having the railroad pass through the town of Lompoc on its way to Santa Barbara, when, should that route be decided upon, the town may hope to reach a population of thousands and become one of the busiest centers in all the western portion of Santa Barbara county, for then the adjacent lands adapted to general farming, fruit and dairying would soon come into market and find ready and willing takers at enhanced prices.

Few have any idea of the mammoth ranchos lying within a radius of twenty miles about the Lompoc Colony, and which, with railroad facilities connecting the colony with Santa Barbara and Port Harford, would converge their trade and business to that point. The most valuable of the properties are as follows: San Julian, Cojo, La Espada, Sal Si Puedes, Santa Rita, Santa Rosa, Purisima, and the Jesus Maria. [For number of acres in each see page 12.] Of all these valuable properties only one, the Santa Rita rancho, is offered for sale and settlement in homestead tracts. This valuable property lies immediately east of and adjoining the Lompoc Colony, and is now being sold in tracts from five acres up to six hundred acres to suit the purchaser. These lands embrace some of the finest fruit, grain and grazing farms in the county, and are offered at fair prices.

The lands of all this section are adapted to the production of grain of all kinds, vegetables in great variety and deciduous fruits of the choicest quality. There are many fine small tracts of suburban town property for sale in the town of Lompoc and often good opportunities to purchase farms at a fair advance to cover improvements. This colony is doubtless one of the very best communities in which to rear a family, and it is of such principally that the community consists. The people are fortunate in having a local paper, the Lompoc Record, that has ever defended its principles and advocated its best interests in every particular. By consulting its columns the seeker for a home in the colony can be informed of everything of local interest.

The town of Lompoc is nine miles inland from the ocean, but its lands and settlement extend to the beach, affording a varied scenery and very pretty and attractive drives. The people are happy, prosperous and contented, and should be, for no portion of our great State is blessed with greater natural advantages.

CHAPTER XII.

SANTA RITA COLONY—RANCHOS SAN JULIAN, SANTA ROSA AND JESUS MARIA.

THE broad plains and rolling lands that form the lower valley of the Santa Ynez, and which, according to the table on page 12, are understood as the Lompoc district, have been hitherto the particular strongholds of the sheep interest. In 1874, the firm of Hollister & Dibblees owned a principality of 140,000 acres of land in this county alone, having an ocean frontage of more than twenty miles and all this devoted to sheep. In those days the traveler might ride for miles and miles, seeing no glimpse of human life except an occasional lonely herder or vaquero. But one by one, the vast ranchos are being partitioned into smaller domains, until upon the plains once sacred to roving bands of sheep may now be noticed fields of golden grain, with cosy homesteads and neat school houses at intervals.

The Santa Rita colony is a community of about fifty families, located on public lands found vacant between the Santa Rita and Santa Rosa ranchos. It lies eastward from Lompoc, eight miles. The temperance sentiments of Lompoc also pervade this colony, it being principally made up of families who were at one time residents of Lompoc. The community is a prosperous one. Since the title (which had been in dispute) was confirmed to the settlers, they all feel wealthy on account of the enhanced value of their property. The land is similar to that directly around Lompoc, of the same quality and productiveness, and with a climate which is, if anything, somewhat superior. Fine homes can be purchased here in 160-acre tracts, at a very moderate price, considering the soil, climate and desirable surroundings. The natural outlet of this community is Lompoc; and when there shall be good roads between these two colonies and a proper bridge to span the Santa Ynez river (which sometimes rises to unseemly heights in the rainy season) no other place would be apt to be sought for business intercourse.

The broad acreage of the San Julian is still devoted to grazing. This is the largest rancho in this district and lies southerly from the Santa Ynez river. Its magnificent extent of fine rolling lands, hill and bottom lands, well watered and timbered, as yet untouched by the slaves of Ceres. The Santa Rosa rancho, north of the river and just west of the San Carlos de Jonata, is also a magnificent property, belonging to J. W. Cooper. It too, is well timbered, well watered, and with fine tracts of bottom lands, and broad, rolling lands, which would, if divided and settled, make a number of comfortable farms. A large portion of this rancho is especially fitted for the dairying business.

The Jesus Maria rancho, owned by J. Ben Burton, is washed upon its western border by the Pacific Ocean. Perhaps it might more properly be classed with the ranchos of the Los Alamos valley, although the Santa Ynez river flows along its southern boundary; but the Los Alamos river, called here, however, the San Antonio, runs through the northern portion of the rancho. It would well make two principalities, and thus be classed in both valleys, and this is the way in which the question was decided on page 12. The San Antonio creek is noted for being the most southern stream which is ascended by salmon in the season. They come into this stream every year from the sea and are caught there—not however, in great numbers. Some 4,000 acres of the Jesus Maria have been put in grain this year, mostly wheat. All grains succeed well. Much of the land is rented out; some of it to dairymen. With its two rivers and numerous springs, the rancho seems especially adapted to this business. In fact this might be said of a large majority of the lands bordering both on the upper and on the lower Santa Ynez.

CHAPTER XIII.

LOS ALAMOS.

THE third natural division of Santa Barbara county, the valley of Los Alamos, lies principally inland, its larger portion, containing the town of Los Alamos, being twenty-five or more miles from the coast. It has been described as a long, narrow valley, lying between the Santa Ynez and the Santa Maria valleys. It is watered by the Los Alamos arroyo, which rises in the San Rafael mountains, and empties into the Pacific Ocean, after traversing the county for about forty miles. But although the valley of Los Alamos is classed as a "narrow" valley, there are large areas of rolling lands on each side, which are rich and productive, not to mention the fertile valleys between, opening into it. The soil of the valley itself is of rich adobe loam or sandy loam, and seems to be especially adapted to wheat-growing. It is estimated that in the district of Los Alamos, (see page 12) there are 40,000 acres of the finest agricultural land. Although the valley is ostensibly drained by the arroyo, yet as it occasionally disappears from sight, as noted in a previous chapter, wells become necessary for domestic purposes. Yet water is always obtainable at a moderate depth, and there is no necessity for irrigation, except in young orchards.

The name Los Alamos means "the cottonwoods," and there were once many of these trees upon the ranch. The following description of the town is taken from the Los Alamos Herald, with the kind permission of the writer, S. R. I. Sturgeon, Esq.:

In the year 1867 Mr. John S. Bell obtained by purchase from José Antonio De La Guerra y Carrillo, that portion of the rancho on which the town of Los Alamos is situated, and for ten years devoted it to the raising of sheep and cattle, laboring all that time under the idea prevalent among old Californians, that nothing but grass would grow without irrigation. This state of things lasted until 1876, he having in the meantime in 1869 built the house in which he now resides, since which time his residence has been in the vicinity of this place. In 1873 the stage route, which previous to that year had passed through the Tinequaic or Foxen Ranch, was changed to the road it now travels, and through La Graciosa to Guadalupe, and the building now occupied by L. Kahn as a restaurant, was erected for the purpose of furnishing meals to the passengers by stage, and also a barn to furnish accommodation to the stage horses. In 1876 the first attempt at farming was made by C. D. Patterson, and proved such a great success that the future of the valley as a producing center, was assured, and seeing this result in the near future, Mr. A. Leslie, still one of our most enterprising merchants, built a residence and opened a store for the benefit of the future farmers of the vicinity, followed by Mr. Snyder, proprietor of the Union Hotel building and the Los Alamos Sample Rooms, which has since grown into one of the best hotels in this end of Santa Barbara county.

In 1878 Mr. Patterson built the first livery, feed and sale stable and has since found it so profitable that he has kept adding to it year by year, until at the present time he is not only doing a fine business in that line but has accumulated sufficient to own one of the finest farms in the valley, as an example of what can be done by a capital of energy and perseverance. In 1877, Mr. Bell, in conjunction with Dr. J. B. Shaw, who had become owner of a portion of the Los Alamos rancho laid out the town of Los Alamos, which now has one hundred dwelling houses and all occupied.

In 1878, Mr. Bell, to further insure the prosperity of the new town and its immediate vicinity, built a fine steam flouring mill, which since that time has been run under the management of John A. Purkiss, one of the best flour makers in Southern California, and who has so far been able to supply the surrounding country with all the flour needed, without as yet ascertaining the full capacity for production of his mill. In 1882, Mr. Peter Coiner built the first public hall in the place, which is now occupied by the I. O. G. T. The lower floor is used as a justice's court room and carriage painting establishment.

We have a nice school house on land donated by Dr. Shaw, for that purpose.

We have not as yet a court house but we have a branch jail, which, if not ornamental, is at least useful, as therein we give lodging and meals to all prisoners from Santa Maria on their way to the Hotel Broughton, and are equally willing to accommodate our friends from Lompoc, only they are so very temperate in all things, that it would appear as if they had no candidates for our hospitality.

Of merchants, we have three dealing in general merchandise—A. Leslie, Laughlin Bros., and A. Weill & Co., from whom or through whom you can procure anything from a needle to an anchor, principally the needle. Also one dealer in watches, clocks, jewelry, Yankee notions, canned goods, fruit and a variety of articles, from shaving your pocket to shaving your face, as in addition to all the articles they have for sale, one of the partners sells his skill as a barber.

We have two blacksmith and wagon-making shops where all kinds of repairing of agricultural machinery is done, or new work built, and one in which horseshoeing is a specialty.

About fifteen months ago the Pacific Coast R. R. reached this place, where they have erected a fine passenger and freight depot, water tanks and all the buildings necessary for carrying on their business; also a telegraph line, putting us into telegraphic communication with the outside world. For the last year we have had a lumber yard with a sufficient supply and assortment of lumber for the immediate neighborhood.

And now if we had a good hardware establishment, we would be ready to enter into competition with any of the other towns in our end of the county."

In addition to the points noted by Mr. Sturgeon, we must add that the town is also furnished with a lawyer, (S. R. I. Sturgeon, Esq.) and a newspaper, the Los Alamos Herald. Religious services are provided by the M. E. Church, the minister in charge of the circuit supplying also Santa Rita, Ballard's and Santa Ynez. Of Societies, the Knights of Pythias, A. O. U. W., I. O. G. T. and Chosen Friends are represented.

Part of the Los Alamos and the adjoining rancho of La Laguna, are owned by Dr. J. B. Shaw, whose residence is in the city of Santa Barbara. In 1867 he bought three leagues of La Laguna, of Celadonio Gutierrez and two leagues of Los Alamos of José Antonio and Antonio Maria de la Guerra. This was a few months previous to Mr. Bell's purchase. The town was built upon lands belonging to both. A large portion of Dr. Shaw's rancho is under cultivation, partly by himself and partly by tenants settled upon it. Mr. Bell's lands are also farmed to quite an extent, but some portions are devoted to sheep and stock-raising generally.

The wheat grown in this valley and ground in the flouring mill by Mr. Purkiss, is said to be of so fine a quality and so sweet to the taste, that the children of the valley cry for it in preference to the finest cake. For this we have the testimony of Dr. Shaw himself.

Next the Los Alamos rancho is the Todos Santos, belonging to the Newhall estate. The Casmalia rancho, on the coast, just north of the Jesus Maria, was purchased two years ago of Mr. Burton, by Merritt & Phœnix. It is a fine property, rolling lands, used for sheep and dairies. It is well adapted for cutting up into small farms, being well watered and with excellent soil.

CHAPTER XIV.

SANTA MARIA VALLEY.

WE are indebted to Mr. James Morse for the following description of the Santa Maria Valley, with its two settlements, Santa Maria and Guadalupe:

Santa Maria Valley, as generally understood, is almost wholly in Santa Barbara county. A small portion of it called Oso Flaco, bordering on the ocean, is in San Luis Obispo county; but as Oso Flaco is a part of the Guadalupe rancho, which lies mainly in Santa Barbara county, I think I may as well include it in my description of this portion of Santa Barbara county.

Santa Maria valley runs due east from the ocean and is about twenty-five miles long by an average of five miles wide (not including the sandy plains on the south side). East of this valley and connecting with it at Fugler's Point is the valley of the Sisquoc, which continues due east some ten miles and is two miles wide in the widest part. On the south side of the valley and about four miles from the ocean is a beautiful lake, which is about three miles long by half a mile wide. This lake is a great resort for ducks and wild geese; white swans also frequent this lake in the winter.

Nearly every acre of land in this valley is good agricultural land. The Guadalupe rancho, containing over 32,000 acres, covers the west end of the valley. As a cloud of litigation has hung over this and the adjoining rancho on the east, there has never been a permanent, healthful settlement. The land is mostly a rich black loam and adobe mixed, and in favorable years has produced enormous crops of barley, which seems to be the crop best suited to the soil and climate. Some years wheat does well but is not so sure as barley. It has been proven that much of the soil is peculiarly adapted to beets for the purpose of the manufacture of sugar. It is also well suited to the production of squashes, beans and potatoes. Mr. T. S. Brown raised a Peerless potato last year that weighed seven and one-quarter pounds! Fruit does well if protected by wind-breaks. Mr. Brown also raised a gloria mundi apple the same year that weighed nineteen ounces and an

Alexandria, seventeen ounces! He also had good cherries. He has an excellent wind-break of cypress. This rancho is well watered, there being several small never-failing streams of water and several sloughs. Plenty of water is obtained at a depth of ten to fifteen feet but is not of good quality. At a depth of one hundred feet very excellent water is struck, which rises in the pipes to within six or eight feet of the surface. There are several flowing wells on low ground. On the Oso Flaco are hundreds of acres of willow monte, and on the hills on the south is an abundance of oak wood. The west end of the rancho, for about four miles from the ocean, is almost entirely devoted to dairying. There are at least twenty-five dairies. These are mostly owned by Swiss, and all are engaged in making butter, only one dairy being devoted to cheese-making. From sixty to three hundred cows are milked on these dairy farms, and it is a very profitable business.

The town of Guadalupe was mostly built up in the years 1872 and 1873, and has been a very busy town, doing nearly all the business of the whole valley until the last two or three years, since which time Santa Maria has gained the ascendancy.

The Laguna rancho joins the Guadalupe on the east and is a fine body of land. Many hundred acres have never yet been plowed.

East of the Laguna rancho is the Santa Maria settlement. This settlement was made in the year 1869 on a large tract of Government land containing about 60 square miles—sufficient for 240 farms. During the first three years these settlers had very hard times. Under the most favorable circumstances the pioneers in a new country have to experience many deprivations and hardships, but these settlers suffered much more than the average pioneer. From the first they had to contest their claims in the courts against the claimants of the Laguna and Guadalupe ranchos, and besides the grasshoppers and wild horses destroyed almost their entire crops for three years. They had to haul all their lumber from Port Harford, about thirty miles distant, and pay for rough lumber there $30 per thousand feet. Some of these settlers had to take their horses into the mountains to keep them from starving, and kill deer and dry the venison, and gather honey to sell in order to buy food and clothing for their families.

In 1868 Pat O'Neil started a store at La Graciosa where a few families had settled. In 1870 Wm. L. Adams started a similar store about two miles west of the present town of Santa Maria. Though he had but small capital, he had good backers in San Francisco and through them and good financiering he was enabled to greatly assist the settlers through these first years of their distress. Here he still stands behind the counter in the old dilapidated store, regardless of the fact that a busy and flourishing town has sprung up less than two miles away. And many of his old customers, still grateful for the many great favors he has shown them in their time of need, drive through town and go to him for their goods. He has built up a beautiful home, with orchard and vineyard, and eucalyptus wind-breaks; is the possessor of thousands of acres of good land, many thousand sheep, cattle and hogs. He has earned it all through honest, fair dealing, and enjoys the esteem of all who know him.

The price of land has doubled during the last two or three years, and improved farms now sell at $40 per acre; unimproved from $8 to $20. The experimental period has passed. The people of this valley are now very prosperous and contented (many are not contented because they cannot get hold of all the land they want). It is proven that fruit will do well here, and in a few years a cannery and many drying houses will be needed. A railroad running through the valley to Port Harford satisfies a long felt want. The climate is very healthful and is every year getting more agreeable; society is good; schools, churches and fraternal orders are plentiful, and a people thus blessed ought to happy and contented.

The town of Santa Maria, located 10 miles east from Guadalupe has a post and express office, a weekly paper, (the Times) two hotels, several stores, blacksmiths, wheelwrights, etc. East of the valley proper are many beautiful, and very fertile small valleys and cañons, and elevated plains and rolling hills, sufficient to support hundreds of families and also a prosperous town. And it is only a question of a few years when a town will grow up somewhere in the neighborhood. There are already three stores, a blacksmith and wheelwright shop, and a church scattered through that section. It is supposed there is considerable good government land yet in this end of the county, but if there is it is all claimed by grant-holders.

Before the railroad was built Point Sal and Morritto were the principal points of shipment for the products of this section. Point Sal is now dismantled, but Morritto's wire suspension chute landing is still running. About one and one-half miles from Morritto is a gypsum quarry of very excellent quality which is being steadily mined and shipped to San Francisco.

Along the coast of this valley gold is found in the black sands of the beach in quantities that pays well at certain seasons of the year to wash. Near Mussel Rocks, Abernethy and St. Ores washed these sands with success for three or four seasons. Now the claim is owned and worked by Abernethy and Wall. Three men often wash out $150 per week. Santa Maria is about 30 miles by rail south from San Luis Obispo, and 40 from Port Harford, and about 80 northwest from Santa Barbara.

OBJECTS OF INTEREST.

POINTS VISITED BY TOURISTS IN AND ABOUT SANTA BARBARA.

Punta del Castillo—The Arlington—The Old Mission.

A VERY good idea of three salient points in the city of Santa Barbara is given in the above picture. First, the Punta del Castillo, showing the curve on the beach between Stearns' wharf and the rocky point, with the slow sea crawling up the sand, the marks of carriage-wheels, and upon the bluff the mansion and grounds of Mr. Dibblee. This is always a favorite spot for idle tourists. The walk from the wharf to the point—about half a mile—just serving to give one a good appetite for his dinner. A road was once built through the rocks, that carriages might drive beyond the point, where a second curve of beach lies below the cliffs; but it was found impracticable to keep it in order, owing to a peculiar fancy of the waves for tossing things carelessly about, and although considerable money had been spent upon it by a public-spirited lady, the attempt was finally abandoned. The road-making changed the point in some respects, and obliterated a shelf known as the "lovers' seat," where formerly one could sit above the surf upon the seaward side of the rock, wholly hidden from strollers along the beach.

The centre of the illustration gives a view of the Arlington, a structure which is a source of pride to every citizen of Santa Barbara. The picture, however, fails to portray the beautiful grounds surrounding the hotel and covering five acres. It gives no idea of the always emerald lawns, dotted with palms and other sightly growths. of the gay beds of choice flowers, the circle of lilies about the fountain, the tennis-lawn and deer park, or the thickly-growing groves in the back-ground. The Arlington is the most capacious and finely appointed hotel in Southern California, designed especially for first-class patronage. Besides its handsome suites of rooms, one of which is called "the Princess suite," having been occupied by the Princess Louise, there are ample parlors and wide airy halls; and from the broad verandas filled in the season with chattering groups of tourists from morn till night, a magnificent view is presented of the mountains and ocean with the distant islands lying like amethysts upon the blue horizon.

Lastly, the striking Moorish architecture of the old Mission is shown, with its long façade and two white towers, though it appears in the picture far less prominent and picturesque than it is in reality. It was built of sandstone from the neighboring hills, with walls over five feet in thickness and supported by heavy buttresses. In front of the massive edifice still remain the ruins of a large fountain of ornate workmanship. The church is high and narrow, 150 feet long, with an organ loft at one end and the high altar at the other. In the vault beneath lies the remains of the first Bishop of the two Californias, Don Francisco Garcia Diego, and over his tomb still hangs the prelate's antique sombrero. The vault was recently re-opened for the interment of the good old Father Sanchez, who had ministered for years at the altar. The walls inside the church are covered with paintings, some of a most eccentric character. In the belfries of solid masonry still swing the ancient bells; although one which had become useless by reason of a crack in its mighty sides, was recently recast and returned to the Mission, as good as new, and can once more mingle with the somewhat harsh and dissonant chime. To the left of the church is a wing, one hundred and thirty feet in length, which with the pillars and arches of its front corridor is still in fair preservation and inhabited by what remains of the grey friars of St. Francis, being the only monastery in the United States. The good Padres have in their possession many of the ancient archives of the Missions, and lists of the flocks and herds of by-gone days, invaluable to a historian, and which Mr. H. H. Bancroft, in his grand historical work, has noted as the most complete of any to which he had access. On one side is the large old olive orchard and all about are the remains of the old dwellings of the Indian converts, though much has been displaced in the last ten years.

It is well known that the Fathers selected the choicest sites along the coast for their own use; it is evident they were men of sound judgment and clear sighted prophecy in many respects, and cultivated tastes. In the location of this Mission they were particularly happy; building stone, fuel and timber were abundant and an inexhaustible supply of mountain water close at hand; they were sufficiently removed from the sea to be secure from hostile attacks from that quarter with such naval ordnance as was then in vogue, and on an eminence which swept the valley east and west for many miles; its white domes being the first object to meet the eye of the traveler from whatever direction he might approach; the position was commanding, the soil rich and kindly, the scenery unsurpassed. The keen-eyed Padres had marked the place on their frequent trips along the coast, both by land and sea, during the seventeen years they had been in the country. But we do not propose here to give a history of the Mission; our province is but to point out to the traveler the sights of Santa Barbara, and of these the Mission is not the least worthy of a pilgrimage.

SURF BATHING.

Between the Punta del Castillo and Stearns' wharf, about half a mile of beach is the favorite resort of bathers. Above all other attractions, the possession of this beach and the facilities it offers for a dip into the surf, is the peculiar advantage enjoyed by Santa Barbara over all other cities of Southern California. A glance at the tables of temperature will show its superiority to Santa Cruz, the only bathing-place which rivals it upon this coast. In the South it can have no rival. Los Angeles and San Bernardino are inland, San Diego lacks the ocean beach. There is a pleasant superstition abroad, which has been encouraged by careless writers, that sea-bathing is possible at any season of the year. It may be, for a strong and healthy swimmer—and indeed we cannot deny that some enthusiasts are ready to tempt the waves every month in the year. But we would not advise a delicate invalid to wade into the breakers until the beach has been properly warmed by the summer sun. Aside from the half-mile curve alluded to, the level strip of sand reaches away beyond the wharf; and when the tide is low enough to round the points, it is a most delightful drive, even as far as the rocks of the Rincon. This drive to Carpinteria at low tide, returning by the road through Montecito, is a favorite device for killing time on a summer —or winter—day. We give the following account of surf-bathing, written by one of the somewhat-too-enthusiastic gentlemen alluded to, only premising that there is much of actual truth in the rose-colored vision :

"Santa Barbara at the present time stands preëminent for sea-bathing facilities over any other sea coast town of California. The water is of so mild a nature that an ordinarily healthy person can enjoy a bath in the sea every month in the year. Sheltered as the harbor is on all sides, it is a rare exception for the sea to be rough or the breakers at all high. Twenty-nine days out of the thirty the sea is as smooth as a mill-pond, and the most timid bather need have no fear of battling with the surf. There is no undertow. The slope of the beach is so gradual that the bather can wade out for several yards without getting above the waist in the water. The beach itself is as smooth as a billiard table, and is composed of sand of the finest quality. The highest temperature of the water in the harbor has been 66 degrees in the middle of summer, and in the winter the temperature has never fallen below 59 degrees, and has only reached that on rare occasions. The average temperature is estimated at 62 degrees. Santa Barbara is essentially the only place on the coast where winter bathing can be indulged in with comfort. Visitors who have tested the fact are unanimous in their expressions of delight and wonder at the remarkably pleasant and even temperature of the water at a time of year in which they have been accustomed to dread the application of cold water to their skins."

Bath-houses and accommodations in the matter of bathing-suits, towels, etc., have been for the past few years managed by Mr. and Mrs. Fred. Forbush. Their establishment is always open during the season, and in the winter upon stated days or by appointment.

BURTON MOUND.

Many of the residents of Santa Barbara know this interesting spot only as the late residence of Don Luis Burton, as a beautiful shady spot for picnics, and as the destined site of a grand sea-side hotel. Travelers upon the decks of passing steamers admire the beauty of the place, which stands, a romantic landmark of the past, only a few hundred feet from the landing-place. It is a mound, circular in form, standing prominently above the level of the surrounding plain, about 400 feet from the surf which breaks upon the smooth and sandy beach at its foot. The top of the mound is about thirty feet above high water and the mound itself comprises about two acres, although the property of which it is a part contains thirty acres. From the level summit may be seen the shore line for thirty miles or more to the east; to the south, the channel and its towering islands present a fascinating prospect; to the west, the light-house, perched upon the bluff, and nearer, the "Castle Rock" of the Punta del Castillo, around which the restless waves invoke a ceaseless melody. Landward, the city, the foothills, in gold or in green, and the Mission towers combine to form an almost unparalleled picture, and one generally neglected by visitors.

Some years ago this mound with its adjacent surrounding property was purchased by a number of the prominent citizens of Santa Barbara, organized and incorporated as the Seaside Hotel Association. It is held by this association for the purpose of using it as a grand sanitarium and sea-side hotel site. It is the best place for a sea-side hotel in the United States. It has a water front upon a smooth sea-beach; magnificent view in all directions, an abundance of shade trees, old and beautiful, rich soil and a luxuriant growth of green grass transplanted years ago, and propagated from the Sandwich Islands. This grass is remarkable for its softness and its luxuriance.

The mound appears to have been a system of subterranean water courses. Springs flow in all directions, and the most remarkable feature about it is their variety. At one place there is a clear blue spring of sulphur water bubbling up and discharging into the grass beneath the olive groves. At another place an "Iron Spring," the water of which is strongly impregnated and the surroundings covered with iron rust. Near the summit a spring of pure water which is used to irrigate an immense vegetable garden, from which Santa Barbara draws its principal supply of vegetables. The property is intersected or traversed by a stream of water from the source of which the city derives its water supply above the Mission. The water of the sulphur spring is similar to that of the Montecito Hot Springs, except in its temperature. But arrangements are made for warming the water, and baths can now be obtained by invalids. The mound is a favorite place for celebrations. On the "Glorious Fourth" and other occasions, the good people of the city, with their wives and families, throng to this cool retreat, with baskets and orators and all the other impedimenta of the day, making the corridors of the old mansion ring with modern life, which must astonish the historic and prehistoric gentlemen underneath. The following extracts from an article in the Daily Independent of Oct. 19th, 1883, give a vivid description of the traditions of the mound: "For many years the coast of California and Oregon has been explored for ethnological relics. It has been dug up by different experts seeking to obtain the various implements of household goods and gods buried with the dead, who knew the patient labor of the Indian during life passed with him to the grave. In other words, the result of his work did not, as with us, go to the living—that it was superstition, no one in these days doubts. And hence we find in the grave the cooking utensils, the arrows, fish hooks, the crude pan for baking purposes, the tasteful olla for boiling, the flint mortar for grinding corn and beans or seed, and various other implements, the present generation cannot understand for what purpose they were made. Even the everlasting pipe is found buried with the smoker. But speaking of the Burton Mound, its origin is unknown to men now living, but it is known to have been formed of the bones, the trinkets, the cooking utensils and weapons of thousands of natives of this coast. It is in fact one grand catacomb or deposit of human bodies covered with immense quantities of sea shells. The interior of the mound has never been explored. No defiling spade or shovel has been permitted to unearth the immense quantities of Indian remains and relics therein deposited. Sometimes when a tree has died and it has been deemed desirable to remove the stump or roots, in digging it out, the earth was found full of Indian relics such as stone utensils, skulls and ingeniously made articles of ornament. Many efforts have been made to obtain permission to explore the interior of this mound, but thanks to the vigilance and care of Captain William E. Greenwell, a manager of the "Sea Side Hotel Association," the valuable ethnological treasures of the mound remain intact. They are perhaps the most complete and valuable collection of aboriginal relics in the United States and will some day be regarded with more interest than at present.

There is a tradition extant which says that this mound was the regal residence of the Grand Sachem or Inca of all the tribes of this Southern coast. Around its base the Supreme Chief of all the Southern tribes held regal court. Upon it the priests and medicine men of the tribes held their mystic conclaves, and no doubt enacted savage tragedies in centuries gone by.

Vancouver, the English explorer, in his three volumes published in 1798, speaks of this mound as the abode of the Great Chief, which undoubtedly it was; in the year 1883, or 95 years since his visit, it is yet unexplored, and is covered with luxuriant vegetation and embowered with vines and fruit trees.

Macgregor in his three volumes, "Progress in America," published in 1847, speaks of this mound. It is certainly an interesting spot and well worth the consideration of the directors of the various universities throughout the world who might seek to obtain the buried relics of a past race."

HOT SPRINGS OF MONTECITO.

The Montecito Hot Springs have for years past been noted for their healing powers. Passing the "grapevine property" and the beautiful places of Col. Hayne and Mr. Bond, the road leads up the mountain, about three-quarters of a mile of steep ascent, until the spot is reached where twenty hot springs flow from crevices in the solid ledges of rock which form the head of the cañon. And an uncanny cañon it is too! But the sick drink of the waters and are healed. Many dip into the steaming pool for the pure pleasure of the bath. These remedial waters are about 1,450 feet above the sea, and about six miles from Santa Barbara. The various springs have a temperature of from 60 to 122 degrees Fahr. The analysis made by Dr. Oscar Loew shows the following constituents of the two principal springs: No. 1—Sodium carbonate 29.6, sodium chloride 8.7, sodium sulphate 5, silicic acid 4.2, with traces of calcium potassium, sulphuretted hydrogen and free carbonic acid. The other is similar, save that it has but a trace of sodium sulphate, less of the carbonate and chloride and a larger proportion of silicic acid.

The discovery of the springs by Mr. Wilbur Curtiss, who lived there for many years, was made in 1855. His health had broken down in the mines and he was wandering through the country in search of health and upon some other unknown enterprise, when he came across a party of Indians at the mouth of this cañon. One among them was supposed to be over a hundred years old; and he led Mr. Curtiss to these springs, and by signs intimated that he would grow well and strong by bathing in the waters. Through an interpreter, the old Indian then told Mr. Curtiss that since he had been a little boy he had drank and bathed there, and that the springs were "the best in the world." Mr. Curtiss remained, drank and bathed in this pool of Siloam and was healed. He then took up a claim there, seeing that the spot might prove a valuable property in the future. At first, there was but a difficult trail to lead to the springs, and over this, for years he carried his provisions and building materials. This was widened and improved, until there is a fair road, over which a stage carries invalids and tourists and the thousand and one items to make them comfortable. Mr. Curtiss first camped with his blankets, then a tent arose, then the first rude hut, then the little cottage, and so building after building was added, until now it seems like a little mountain village, but alas! the poor old gentleman is no longer of the earth to see the gradual fulfillment of his cherished plans.

It is learned upon good authority, that when the country still owed allegiance to the Crown of Spain, the Spanish government sent out a commission of scientific observers, with orders to make an examination and analysis of all the then known mineral waters in Mexico and the Californias. This commission after spending much time in the prosecution of their inquiries, reported the most favorably upon the properties of the Montecito springs. For rheumatism and scrofulous diseases they are said to prove in many cases a specific; but no invalid should venture to drink or bathe in the waters without medical advice, since the springs are not effective in all diseases.

Vistors to the Springs are always shown the view from "Lookout Point" which is reached by a winding trail on the mountain side. One of the finest views is there obtained, of the valley of Santa Barbara, with its city and suburbs, and the ocean and islands beyond.

These springs of Montecito are the best known and most accessible, but not by any means the only mineral springs in Santa Barbara county. Over the San Marcos Pass other springs have been found and utilized, and there are springs also in many parts of the Santa Ynez range, both cold and hot, generally strongly impregnated with sulphur. The following description of San Marcos or Mountain Glen springs was written by Professor G. R. Crotch: "These springs are conspicuous among the natural beauties of Santa Barbara. The drive over the mountain is in itself replete with beauty; after you cross the summit of the Santa Ynez where all is luxuriant and green, streams crossing the road freely, level plateaus of abundant grass, undulating groves of oak, the chaparral thick with flowers, all form a picture of rare beauty. The springs are situated in Mountain Glen, one of the many cañons crossed by the stage road, and debouching into the Santa Ynez river about six miles from "Pat's," (on the summit) at the San Marcos house, an old ruined adobe building. From here a tolerable road has been cut through the brush for about a mile, the latter part passing through a dense forest growth. At last a little log cabin is reached, and this marks the site of the principal springs. The scene is picturesque in the extreme, a densely wooded cañon, with sides nearly vertical, about 250 feet high and clothed with brushwood; the bed of the stream thickly covered with various forest trees. Two sycamores opposite the cabin are at least 70 feet high and as straight as palm trees. The stream itself, steaming somewhat in the cool air, combines to render the effect unique. The bed of this stream is of cold and beautifully clear water; in this at intervals, arise five or six different hot springs, none too warm for the hands and all slightly impregnated with sulphur. Other cold springs more strongly impregnated, arise also at the edge of the main bed. The warmer springs have coated the stones with a characteristic green coating as the singular springs in the Azores do."

These springs have been found beneficial in many instances, and there are other attractions also in the vicinity. The valley below abounds in quail, the streams are stocked with trout, and if the sojourner be a naturalist a wide and productive field is here open to him, as insects of many interesting varieties have been found in this neighborhood by visitors fond of the "science of bugs and things." The springs are now owned by T. H. Hough, who has made some improvements, and provides very comfortable beds for guests, of whom he has many during the summer months. The stage passes the mouth of the cañon every day, and visitors are there met by Mr. Hough if proper notice is given, or if the guests are not of the invalid variety, they can very easily make their way to Mountain Glen on foot.

THE LIGHTHOUSE.

A short drive, fit for an invalid on a pleasant day, is the way to the lighthouse upon the mesa. The idea of a lighthouse ordinarily suggests a storm-beaten coast, a leaden sky, with the howling winds and furious rain creating a wild uproar, in unison with the savage breakers dashing on the rocky base of a tall and lonely tower. But this is a peaceful and home-like spot. The sky is blue, the blue channel smiles in the sun, unruffled for months at a time. Sometimes a soft haze half obscures the islands; or a fog bank rises from the ocean like a distant range of mountains.

The lighthouse stands upon a fertile mesa, only a half hour's ride from the city. The keeper, who has trimmed the lamp for twenty-eight years, came with her husband when the lighthouse was built. She subsequently assumed the sole charge, and for nineteen years has but once left the beacon light. Mrs. Williams is, in fact, the veteran lighthouse keeper of the coast, and is always ready and willing to exhibit her charge to visitors. The lighthouse is a homelike edifice in its design giving no evidence of the extreme solidity of its walls. The house is built upon most solid foundations, its walls being of stone two feet in thickness. From these rises a tower of bricks inside of which a winding staircase leads to a small apartment near the top. From this a short iron ladder leads through a small man hole in the floor of the lantern room above, which aperture is so small as to require considerable effort on the part of a portly person to squeeze through. On the first landing a notice to visitors, provided by the Government, cautions them against scratching or defacing the lantern or the property in any way. The lamp is a Hains, 32 candle power, visible twenty miles on a clear night, and is a fixed white light. The glass consists of a focal plane with eight prisms; five above and three below. As Mrs. Williams carefully removes the cloth from around the object upon which she has lavished such assiduous care for many years the brilliancy of its shining sides are almost blinding. The reflector shines like a huge diamond, not a scratch upon its surface nor a speck of dust, testifying in its splendid condition the care it has received. The lamp placed back of this glass is an argand burner of anything but imposing appearance. Everything about the tower is a model of good house-keeping or lighthouse-keeping. In the family rooms below are gathered trophies of sea and shore. Shells, starfish, sea-urchins and the like washed up from the sea at the base of the cliff upon which stands the lighthouse itself. The view from the railed platform about the top of the beacon is superb. The Goleta landing is visible to the right, the islands in front and the wharf of Santa Barbara is hid from sight by the graceful lines of Castle Point.

LAKE FENTON.

About four miles from the city, on the Rancho Positas y la Calera, is a lovely little lake in the centre of a rim of hills, which are covered in early spring with a green sward and dotted with stately oaks. It is a pleasant destination for a short drive, and the name was formally bestowed upon the little lake during the visit of Governor Fenton of New York. The party of sponsors included several prominent men from the East, two or three army officers and a few citizens of Santa Barbara. Mr. Charles Nordhoff alludes to the spot in his work on California, and mentions a picnic there on the 22d of February. The lake was formerly known as Laguna Blanco.

AN INDIAN VILLAGE.

•There are many places of historic interest in this valley; only a few of which can be referred to—some of them date so far back that they are hidden in the mists of antiquity.

At Goleta, six miles west of Santa Barbara, is a place of great interest to antiquarians. It is the former site of one or more Indian villages, whose origin is prehistoric. From this locality, named by them the "Big Bonanza" a corps of Government scientists exhumed about ten tons of relics that were shipped to the Smithsonian Institute at Washington, and finally distributed among the leading scientific museums of the world. When the first white man visited this coast, 342 year's ago, there was a dense population in this valley and numerous villages, which were ruled over by an Indian Princess who lived in this village, then called Ciacut. The dwellings have all disappeared, only the kitchen debris remains to mark the homes of this once populous race. The dancing circle is plainly defined, for so firmly had the soil been beaten down by countless generations of dancers, that vegetation still refuses to grow in the sacred inclosure. Cabrillo says: "They had large public squares in their villages, and an inclosure like a circle, and around this they have many blocks of stone set in the ground, which issue about three palms, and in the middle of the inclosure they have sticks of timber driven into the ground, like masts and very thick, and they have many pictures on the same posts, and we believe that they worship them, for when they dance they caper around the inclosure." Their dancing was accompanied by chanting, slapping of hands, blowing on whistles, beating of drums and rattling of shells filled with pebbles. They danced to propitiate the divinity, they danced when successful in fishing or hunting, they danced at a marriage and at a funeral. The historian says such was the delight with which they took part in their festivities that they often continued dancing day and night, and sometimes entire weeks."

MISSION CANON—SEVEN FALLS.

The Mission creek, or rather the Arroyo Pedrogoso, has several times been mentioned as the souce of the city's water supply, and also for the scenic effects along its sparkling path. The day on which no picnickers are camped under the oaks and sycamores, somewhere between the Mission buildings and the Seven Falls, would be a frigid day indeed, and one not likely to be met with in the summer time. Families often take their simple tents and household stuff and stay for days or even weeks. It is so near to the city that any person of average health can easily walk to a pretty picnic spot. But to the Seven Falls is a climb to which only the strongest and most enthusiastic do not succumb; although the reward is great. There is nothing on earth so lovely as a mountain brook, and of all mountain brooks, this is one of the loveliest—the clearest, the purest, the most bewitching. From the hollow just behind the Mission to the furthest springs, is a succession of beautiful, wild, natural scenery. If it has a fault, it is the superabundance of campers (all with distressing appetites) in the season.

MOUNTAIN TRAILS.

Northward from the city, an old trail crosses the Santa Ynez range, into the upper Santa Ynez valley, descending into the abandoned quicksilver camp of Los Prietos. A description by the Rev. S. R. Weldon, of a trip over the range is here given, which will perhaps tempt a tourist or two to go and do likewise. He says: "We struck the mountain's foot at the mouth of the Cold Spring cañon. For a mile or so it was easy work, till we approached the Cascades, where the brook tumbles down a lofty precipice, perhaps 300 feet high, which blocks the way up the gorge. So following the narrow trail to the left, up we climb, stopping every few rods to rest our panting steeds. At first we almost shrink from looking down into the abyss below us, but we soon get accustomed to it. The sense of security grows upon us, and soon we can carelessly gaze down the steep declivity—a single mis-step and we could not stop for a thousand feet. But our sure-footed animals make no mis-steps. They do not stumble; we feel safe and are safe. Yet it would be a bold man who would dare to ride one of our large "American" horses on these mountain trails; but mustangs and mules are very sure-footed and even ladies learn to ride them on these trails without fear. In two and one-half hours we have reached the crest of the long ridge that for so many miles separates the Santa Ynez from the Santa Barbara valley. I marvel to find it so sharp and narrow, scarcely a level yard of ground anywhere. The prospect is most striking. One hundred and twenty miles of coast line beneath, the beautiful valley with the pretty town—you can almost name the familiar streets and pick out the houses. Then the ocean and the islands! How quietly, how grandly beautiful! But turn around; could anything more wild and desolate be imagined. See, range after range of bare, bleak, verdureless mountains. .Thirty miles of waste! You have heard that deer, bears and mountain lions roam amid these starving sierras, and wonder what they find to eat. But there are hidden cañons and unseen valleys, where are forests and grass; sufficiently wild and barren, but not quite so desolate as it seems.

But now we descend the steep declivity to the north. There are quicksilver mines and hints of more valuable metals; but nobody has yet grown rich from the minerals stored among these rough treasure-houses. Perhaps in some future time they may. In a couple of hours we reach the dwelling of the laborers who work at the quicksilver mine; a free-hearted hospitality offers food and lodging. But we came to camp out, and camp we will. So, after a lunch and rest, we wend our way up the valley to a spot where grass, water and shade may be found, and having staked out our horses, build our fire.

Having eaten our supper and gathered the tall fragrant herbs for our couches, we roll up in our blankets to sleep under the stars. How sweet and how exhilarating is the air of the night in these dry solitudes! * * * Next morning, about half a mile from our camp, a bear appears on the opposite side of a gorge, within easy rifle shot; but the brute seems to know that we do not carry Winchester rifles and so leisurely walks along and in ten minutes disappears in the thick trees at the bottom of the cañon. These are very shy animals, seldom seen, even where they are numerous, so we counted ourselves fortunate to have made his bearship's acquaintance. * * * Returning, at 7 o'clock we reached the summit, and there we saw what has often been described, and its wondrous beauty, so difficult to imagine until seen—the clouds at our feet, resting against the mountain side and stretching far away to the distant horizon, the mountainous islands alone pushing up their peaks above the vast billowy expanse, its beautiful folds, white as the driven snow. * * * Then down the steep mountain side and, when in a couple of hours we reached home, the sun was shining there too."

UPPER SANTA YNEZ.

The preceding description of a mountain trail, it will be seen, lands the traveler amid the unique scenery of the upper Santa Ynez, where in comparatively modern ages, Titanic forces burst through the mountain barriers, letting pass the silver river and showing upon the scarred precipices veins of cinnabar, where the strata were torn violently asunder. Since there are no longer miners to offer "food and lodging" the trip can only be made by one who is able and willing to "rough it." But the hunter will find it a veritable Paradise on earth; for deer are plentiful in the cañons and among the wooded ridges, and in the cool shadows of the rocks lurk the wary mountain trout—luxury of luxuries. Here too, the man who has lost a grizzly can find him.

As for scenery, there are places which rival the Yosemite in grandeur; the former mining camp of Los Prietos is overshadowed by a precipice fifteen hundred feet high; a steep and solid rock, up whose face not even a goat could climb. The effect of this great impenetrable wall by moonlight is weird and solemn; its awful shadow looms above like a perpetual reproach against the littleness of man and the insignificance of all his works. Between the two deserted camps of Los Prietos and Santa Ynez, the river runs through a narrow valley, between mountains red with cinnabar, sometimes under overhanging limestone walls and opening out again into little oak-covered flats. One enormous rock, which scarcely allows the stream to pass, was aptly named by the miners Gibraltar, and near its base the still unexplored cave of Najalayegua opens its mouth. In places great red boulders of cinnabar lie piled upon each other with prodigal recklessness. Near where formerly the upper mine was worked, high up on the mountain side, is a little castle built of cinnabar, held as a fort during one of the many fights between rival claimants to the mines. There are other points, besides the abandoned mines, worth visiting by those who can dispense with the "blessings of what is called civilization" and have breath left for climbing. And in the purple distance the San Rafael range seems to stretch away into infinity, with the mineral secrets of its wild unknown recesses—and all impress the beholder with the glory of this gracious and terrible saint in her splendid solitude.

THE SAN MARCOS PASS.

The toll road over the San Marcos Pass, which is the identical spot where Fremont descended upon the town in 1846, offers attractions to tourists, but this is also of the "roughing it" variety. The view from the summit is superb and there are many spots suitable for camping. The air upon the mountain is, in some states of physical disease, quite beneficial; as when fogs cover the valley, it is clear and bright upon the summit. But to mention one mountain retreat is to mention all, and so we will but cursorily speak of other noted spots which are famous among campers.

GAVIOTA PASS.

Taking the avenue which leads past the Hollister and Cooper ranchos, past the lovely Tecolote and other coast cañons which have been mentioned previously, twenty-five miles away, we reach El Capitan, the usual tarrying place of campers en route for the upper country. The valley here is narrow. The wooded cañon, with its gurgling brook, pushes itself out into the sea. Long lines of trees across the open country meet and mingle with the line of the ocean; a low promontory stands out to the south, making a little cove where a skiff might be beached unhindered by the breakers beyond. By the brook, with its mimic falls and pools, we know El Capitan. From thence, we follow for fifteen miles the line of the coast, where numberless cañons open to the sea, until we reach the Gaviota landing; thence turning away from the coast, where a mountain stream has broken through, we make the Gaviota Pass—a natural chasm about sixty feet wide through the range, and within a mile of the sea. It is a delightful trip in summer, but when the summit of the Pass is reached, the cold wind blowing from the ocean meets the dusty traveler, showing the great difference in atmosphere between the two sides of the Santa Ynez. Spoiled by the soft delicious airs of Santa Barbara, the valley into which we descend seems chilly and unwelcome. Yet, compared with other places, nature deals kindly enough with the dwellers in Lompoc and the coast regions above Point Concepcion.

CASITAS PASS.

Eastward from Santa Barbara, there are a thousand spots to tempt the tourist—passing through the valleys of Montecito and Carpinteria, with their unique and charming cañons leading into the inmost recesses of the mountains, we reach the Casitas Pass, and beyond it the splendid scenery of Ventura county and the gem of Ventura, its far-famed Ojai Valley.

THE OJAI.

It is said that every pretty cottage in this valley is tenanted by some one who has fled hither to escape the horrors of the asthma, and has found relief. There are two excellent hotels in the valley, where invalids are well taken care of by Mr. F. P. Barrows and McKee & Gally. It is scarcely within our province to speak of this valley, of its magnificent scenery or the fishing and hunting there enjoyed. But the majority of tourists sooner or later make it a visit. It is also noted as the spot which immortalized itself by naming the little hamlet in its midst "Nordhoff," in honor of the celebrated journalist.

FLOWERS, FERNS, SEA-MOSSES, ETC.

WRITTEN ESPECIALLY FOR THIS WORK BY RESIDENT SCIENTISTS.

THE FLORA OF SANTA BARBARA.

[By Mrs. R. F. Bingham.]

In passing down the beach, you will be delighted to see the beautiful trailing abronia umbellata, intermingled with the delicate yellow œnothera, often with a background of the shrubby lupine. Follow the road around the cliff, and you will always see the California rose in bloom, together with the white convolvulus, and many other flowers, varying with the time of year. Some of these are not found elsewhere, others growing at the base of the hill in the sand, are also found at a thousand feet elevation, and at intermediate points. On the top of the hill, covering large spaces, are the purple brodæas and the orange California poppy, side by side. Further down, beneath the oaks, are mossy spots in which grow the maiden hair ferns, while in more rocky localities the pellæa and polypodium are found. In spring, if you look very closely, you will find a small plant with delicate creamy blossoms, a representative of the poppy family, which will well repay your search. Later in the season as you approach the main road, you will find the hillside purple with the beautiful Collinsia bicolor, and again the more familiar buttercup will greet you.

Almost anywhere along the stream banks you will find the bright blue lupine, the white cardamine, the finely divided leaved thalictrum, and under the trees, the delicate blue nemophila will peep through the grass, reminding you of the wood violet, which grows here only upon the mountains. Many ferns can be found in these localities, as also many bright, showy flowers.

Often times in open spaces, are large masses, white with the bloom of the eretrichium, which in shape and size resembles the blue forget-me-not. The blue-eyed grass smiles at you everywhere. Purple peas, yellow hosackias of many sizes, purple and white godetias, phacelias of various hues meet your eyes at almost every turn.

Perhaps in threading your way among the bushes, you suddenly notice a perfume that reminds you of something you remember of your childhood, but you can scarcely make out what it is; looking carefully about, you find you are treading upon the fragrant yerba-buena trailing along the ground, hiding its delicate blossoms under its small leaves, reminding you of the trailing arbutus, but in no way resembling it. As you ascend, you find the rose-colored and scarlet gooseberries, which, but for the armament of triple thorns, would be well worth a place upon the lawn.

Then there are many species of California lilac, the mountain mahogany, the California laurel and holly. These shrubs are often evergreen, with vines that produce a profusion of white sprays, followed by their curious prickly fruits, and occasionally a clematis will hang out its snowy blooms, or plumy seeds. On the cliffs along the mountain streams, grow the delicate and rare ferns, mingled with brilliant flowers, and on the level side, the towering fronds of the Woodwardia.

In summer the mountain slopes are covered with the snowy spikes of the yucca, rising many feet above the bayonet-like leaves, and you are often surprised by masses of the bright golden helianthemum. In autumn will be found representatives of the sage family, in white, blue and purple, while underneath them are the scarlet-painted cup, and the low-growing crimson chorizanthe.

The varieties of calachortus and other bulbous plants are exceedingly beautiful. Scarlet mimulus in dry stream beds attracts attention, while the salmon-colored one can be found in rocky places all the year. Yellow flowers in the greatest profusion everywhere, and always, until one almost tires in thinking of the great variety.

As you ascend the mountain heights, new faces greet you, intermingled with those you have met lower down.

There is one modest flower you must not fail to find; look under the shrubbery in low places, and you will be repaid for your search by the small, sweet-scented, blue scatellaria, which you will never tire of collecting.

One might fill page after page and scarcely begin to tell the beauties of our flora; words are powerless for description; it must be seen and loved to be appreciated.

SEA MOSSES.

[By the late Dr. L. N. Dimmick.]

The sojourners by the seaside at Santa Barbara, who are fond of the beautiful and strange in nature, will find a rich treat in the exquisitely beautiful sea mosses that are cast upon the shore by the in-coming waves of the blue Pacific. At their feet they will see these "flowers of the sea," brilliant in their rich and various shades of crimson, green, purple, olive and brown. Some of these when growing under water reflect prismatic colors, and as they sway to and fro with the motion of the waves, they glisten with the richest metalic greens and blue with pink and scarlet, each leaf a rainbow in itself. These sea mosses are eagerly sought for, collected in large quantities, and distributed throughout the United States. They are gathered at all seasons of the year, as there are no icy shores to prevent. The spring and summer are the best times in the year to collect, as a greater number of varieties can then be found, and they are of more luxuriant growth.

They are floated out on cards or paper, pressed until dry and then arranged in books or albums. When they are wanted for making sea moss pictures and landscapes, or for constructing wreaths, or filling baskets, they are first dried on bright tin plates or oiled paper, from which, when dry, they can be readily removed in a fit condition for this highly ornamental work.

The Atlantic coast possesses many lovely varieties, but the Pacific far exceeds it in the number of its species and varieties. The red-colored sea mosses are the most abundant in deep water, and are most likely to be found after a rough sea has loosened them from their attachments. There is a marked difference in form as well as in colors between specimens of the same species when growing at different depths. The sea mosses, or Algæ, are the food of a large number of marine animals. Man too, has found some of them very desirable as articles of diet. The Irish moss of the Atlantic finds here its superior in the Gigartina family, which is free from bitterness and makes a superior jelly. It is possible that the salubrity of sea air is due in part to the Iodine set free from the kelp or dark green sea weeds that dry up on the shore.

MOLLUSCA.

[By Dr. Lorenzo G. Yates.]

California includes within its limits a greater variety of natural regions and peculiar zoological districts than any other State or Territory of the Union. Its seaboard of over 600 miles in length, furnishes a great range of climate, and a consequent varied fauna and flora. Santa Barbara, from its peculiarly sheltered position, caused in a measure by the conformation of the coast, marks the boundary line of the northern and southern limits of many of the marine species of mollusca.

The Pacific Ocean along our shore keeps up an uniformity of temperature suitable to the requirements of many subtropical forms.

Many of our genera are peculiar, the sub-genera of land shells, in particular, and differ from those of the central and eastern portions of the continent, while many of the eastern genera are unknown with us. At the same time the localities where these peculiar genera or species are found are sometimes very limited; for example, *Binneia notabilis*, found only on the Santa Barbara Island, and these confined to the south-east side; still another species, *Patula Durantii*, found in the same locality as the above, and nowhere else except in Alameda county, (300 miles north) where it has been found in limited numbers by the writer and others.

Helix Ayresiana, a very pretty snail shell, found on the Channel Islands only, and which will in all probability soon be exterminated, pasturing of sheep on the islands destroying the herbage on which they feed and under which they are protected.

Within the city limits shells are rarely found of any beauty or scientific value, but above and below it are points where the localities are favorable for the collection of rare and fine species.

Although Santa Barbara has perhaps fewer species than some other points on the coast, still the coast of the county contains a large number of the species found north and south, and some not found elsewhere. It is the northern limit of Luponia spadicea, (a very pretty cowry or money shell) also Siphonalia kelletii, a rare and fine species. Live shells of the above named species are occasionally obtained by dredging in the channel; they live upon the kelp and in deep water. Broken and dead shells are not uncommon on the shore, and doubtless many species not now known in this locality may be found by dredging in the channel.

The islands afford not only a sheltered coast, but also an extraordinary length of coast line, the rocky portion of which furnished the native Indians an abundant supply of food molluscs, while the shells furnished them with ornaments, articles of utility and material for exchange or barter with the interior tribes, while at the present time the Abalone or Haliotis furnishes valuable articles of commerce; the Chinese dry and ship to China immense quantities of the animal; the shells are polished and sold as curios, and large quantities of the rough shells are exported for the manufacture of jewelry, buttons and various other articles. The delicate and beautiful Paper Nautilus (Argonauta Argo) is sometimes found on the islands after a storm; it is highly prized by collectors, and lovers of the rare and beautiful in nature.

Another curious and pretty shell, Lucapina crenulata, was used by the Indians as ornaments. The Bubble Shell, (Haminea vesicula) very delicate and transparent, is sometimes found in the Estero.

Extensive and interesting collections of shells have been made here from time to time by scientific explorers.

In 1866 Mr. Hepburn collected 70 species in Santa Barbara and vicinity. In 1867, Dr. Newcomb, an eminent conchologist, now of Cornell University, collected 135 species on the coast of the main land, and 23 species on Santa Cruz island, some of them quite rare. The writer collected 160 species on Santa Rosa island, a complete list of which was published Nov., 1876, in the Quarterly Journal of Conchology, at Leeds, England. This collection was made incidentally while engaged in searching for antiquities in the interest of the Smithsonian Institution. Doubtless the list could be considerably extended.

The seaward side of the islands being exposed to the full force of the wind and tide, and the land side moderately calm and sheltered, furnish widely different conditions for the various species of deep-water and littoral molluscs and offer extensive opportunities for further research and discoveries.

FERNS.

[By Lorenzo G. Yates.]

The remarks on the peculiarities of the climate of this county as regard the molluscan fauna, will apply also to the flora of this region so far as the Felices or Ferns are concerned. While some of our species are found in the northern part of the State, and are not found south of us, others begin here and extend south. Santa Barbara being the southern limit of species found in northern California and Oregon, and the northern boundary of others found south, into Arizona and Mexico; others again are peculiar to this and the counties immediately adjacent, and one species at least, is peculiar to Santa Barbara county on the Pacific Coast, and found elsewhere only in Florida and Texas. We find the following species in this county: The Adiantums or "Maiden-hair Ferns," are remarkably well represented, all the species known in the United States, with perhaps one exception, being found here.

Adiantum pedatum, in the northern portion of the county. A. capillus veneris, A. emarginatum, in the wooded cañons in the neighborhood of living water; hanging in immense masses from the boulders under the spray of falling water.

Aspidium patens, a remarkably fine and pretty fern; the only localities where it is found in the United States being Bartlett Cañon in this county, and in Florida and Western Texas.

Aspidium rigidum, variety argutum, found throughout the country in abundance.

Cheilanthes Californica, commonly called lace fern, a very delicate and pretty species found only in the Coast Range of California, and much sought after by collectors.

Cheilanthes Cooperæ, first discovered in Santa Barbara county and named for Mrs. Ellwood Cooper, a lady who takes great interest in the collection of ferns.

Gymnogramme triangularis, the golden back or gold fern and under certain conditions of growth taking the form of the so-called silver fern, common everywhere.

Pellæa andromedæfolia, P. ornithopus and brachyptera; all the pellæas improve under cultivation.

Peteris aquilina, var. lanuginosus, brake fern, common.

Woodwardia radicans, variety Americana, is found growing in and near mountain streams. Magnificent specimens of this fern are used to decorate our halls and public buildings on special occasions, notably our Floral Carnival.

INDIAN REMAINS.

[By Dr. L. N. Dimmick.]

Of the inhabitants of this county, previous to its discovery by Cabrillo in 1542, nothing is known except as developed by a minute examination of their rancherias and cemeteries. From these have been obtained many tons of their household utensils, tools, weapons, ornaments, and various other articles that throw light upon their domestic economy, occupations, character and history. When this coast was discovered by Cabrillo, no other portion was found so densely populated as this vicinity. The early records of the Mission give the names of over one hundred and fifty clans or rancherias that were located within the limits of the territory afterward formed into the county of Santa Barbara. The supply of food appears to have been so abundant that there was no struggle for existence, and the climate so even and delightful, that they showed their appreciation of these conditions by crowding it with a dense population, who for a long period enjoyed here a peaceful and indolent life. Excavations into the cemeteries show that many of the localities had been occupied continuously for probably ten centuries at least.

With the skeletons, that from the measure of decay, appeared to have been buried from one hundred to three hundred years, were found a few modern beads and other articles of European manufacture, mingled with stone, wood, bone and shell implements. Still deeper, beneath these graves, were found remains more decayed, with only the stone, bone and shell utensils. Layers were found of deeper and deeper interments, in which the human remains crumbled into dust upon being exposed to the air. These skeletons exhibited an antiquity equally great with the remains of the mound builders of the

Mississippi valley. The skulls resembled those of the more intelligent of the native races. The bones indicated a muscular race, of medium stature, somewhat taller than the inland tribes. The sites of their villages are covered with the remains of mollusks, fish and seals, showing that from these animals they obtained the larger portion of their food. The rarity of warlike implements indicates that they were a peaceful race; their care for the dead proves that they were not destitute of natural affection; and the fact that they buried with their departed friends all the implements and other articles of value belonging to them, testifies that they believed in a future state of existence, where these articles might be required. The bodies were usually buried with the face downward, and the knees drawn up under the body.

With many of the skeletons of females were found balls of red ochre. Sometimes this was carefully preserved in abalone shells, or in small stone cups. Bracelets and necklaces of bone and shells, together with strings of shell beads and shell ear-rings had been buried with them. The most common domestic utensil was the stone mortar and pestle, which were of all sizes, from those holding three or four gallons down to those holding less than a pint. In these they doubtless pounded their acorns and other seeds, which they seasoned with grasshoppers when they were plenty enough. They had tortilla stones cut out of soapstone or steatite that were fire-proof, on which they baked their acorn cakes. They also carved from this same kind of stone neat cooking utensils. These were globular, with rather narrow apertures, often encircled by raised rims, and would hold from half a gallon up to four gallons. Cups, bowls and ladles were carved from serpentine and highly polished. Rude knives and awls were made of flint and bone. Abalone shells were used for drinking purposes and for plates. Needles were made of bone.

Highly polished serpentine pipes, with hollow bone mouthpieces, cemented in place with asphaltum, indicate that they liked to enjoy their ease when smoking, as the straight, elongated pipe was only adapted to be used with comfort when the smoker was in a recumbent position. They made fish-hooks of both bone and shell. Arrow and spear heads were of flint, as were also the scrapers with which they dressed and prepared the sealskins for their clothing.

Remains of nets and the abundance of sinkers found on the islands where the best fishing grounds exist, show that they were experts in this mode of catching fish. These sinkers were generally discoidal stones, with the opening in the centre beveled. It is probable that they had secondary uses for these stone rings, and that they were used in playing games. One form of these discoidal stones is club-head in shape, and is supposed to have been used on sticks of wood for convenience, in digging the ground for roots. Whistles and flutes of hollow bones of birds show that they were not entirely destitute of musical taste. Their shell money was generally composed of small, round pieces of flat shell, perforated in the centre, or else small shells like the olivellas, truncated at the apex so as to be strung together. Beautiful models of boats were carved in serpentine. As the northern tribes around the Sacramento river and the bay of San Francisco knew nothing about boats, having only balsas, (which were small rafts of tules or rushes) the possession of these fine models that they evidently prized highly, and the admirable boats they possessed in abundance when Cabrillo first visited them, and which he describes as constructed of bent planks, cemented with bitumen, the largest of them capable of safely transporting twenty persons at a trip across the channel intervening between the main land and the outlying islands, proves them to have been a much more intelligent race than any of the more northern tribes.

But as soon as the eye of the white man rested upon them they commenced to melt away. A little more than three hundred years later, and the native race was almost extinct. This fair domain, once their exclusive possession, is now in the occupancy of another people who wander over the deserted homes that are all the record this vanished race left of their history; "of their inner life, their aspirations, hopes and fears in the unrecorded past."

[This article is a reprint, given in place of the one promised by the late Dr. Dimmick, who took great interest in this undertaking; to whom in fact, the plan of the work is largely due. He had prepared the papers on "Sea-Mosses," the "Indian dancing ground at Goleta," the "Adaptability of the soil to lemons," and some paragraphs on minor subjects, when he was taken seriously ill; and when his death occurred, the first half of the work was already in press. In his death the city has sustained a severe loss. He was not more noted for his scientific attainments than for his generous heart and liberal public spirit. Even when confined to a sick room his influence was widely felt in all matters pertaining to the welfare of the city; his advice was prized and his judgment relied upon; strangers were referred to him as to the best authority—and not to know Dr. Dimmick was not to know Santa Barbara.]

GEOLOGY AND PALEONTOLOGY.

[By Lorenzo G. Yates.]

At various and widely separated points over the area of the State, California has several noted fossil localities; one of the most widely known being the strip of land occupied by the low-lying hills of Post-Pliocene age, lying between Santa Barbara and the ocean. These hills arise at the highest point to about 433 feet, and are composed of strata of nearly horizontal rocks which rest upon the upturned ledges of the underlying bituminous shale. These rocks consist of coarse gravel and sand, in some places very hard, at others soft and friable.

These strata which have been referred to the Pliocene and Pleisto cene or Post-Pliocene periods by different authors, are again overlaid unconformably by alluvial deposits. At this place the first collection of fossil mollusca of which we have a record, was made in 1849, by the late Col. E. Jewett, who was well known to the people of Santa Barbara, where his daughter and her family still reside. A description of these fossils was published in 1863 in "Reports of the British Association;" many of the species being new to science, and several were named after Col. Jewett. Forty-six species were enumerated, about one-half of which are still found living in the waters adjacent; one of the most noted fossils *Crepidula grandis*, the "boat" or "slipper shell," is at present a boreal species, found living on the shores of Kamtschatka; another species, *Janira bella*, illustrated in Vol. 2, of Paleontology of California, is well known to collectors, and prized by tourists and others, from its peculiar shape and excellent state of preservation.

Extensive collections were also made by the geologists of the Pacific R. R. survey, and are described in Vol. 5, of the "Reports."

Several years since, Rev. S. Bowers and wife collected over one hundred species of fossils in this locality. In the same range of hills, west of the town, large numbers of *Turritella Cooperi* may be found in a good state of preservation. Some eight years ago this fossil was found by the writer in large numbers at a point on the bluffs near More's Landing, weathered out on the surface.

The late Prof. W. M. Gabb, Paleontologist of the California Geological Survey, refers the Santa Barbara fossils to the Post-Pliocene epoch.

In the range of hills lying east and north of the city, extensive deposits of diatomaceous earth or rock may be seen, and in this locality Dr. Finch has discovered the fossil bones of an animal which from a cursory examination, appear to be those of an extinct animal allied to the Dugong or Manatee (sea-cow, a herbiverous, swimming mammal, having the aquatic habits of the whale, and being one of the species on which the fable of the mermaid was founded) remains of which the writer discovered in Alameda county in this State, several years ago, the teeth of which have puzzled the most eminent of Paleontologists and comparative anatomists.

Teeth and bones of the Mastodon were found in this county several years ago by the State Geological Survey.

In 1876, the writer found a fragment of tusk of fossil elephant on Santa Rosa Island, in this county, (a relic of the times when the channel islands formed a portion of the main land).

In Mission cañon and at other points along the line of the Santa Ynez range, are localities of Miocene fossils, principally oysters and pectens, (scallop shells). Good specimens of these fossils may be found in the Gaviota Pass.

Near the last named locality, fine specimens of fossil teeth of sharks have been found on the surface of the ground, having been weathered out of the miocene rocks. At Indian Orchard a tooth of another and interesting species of shark was found, specimens of which are in the writer's collection.

AGRICULTURE.

Cereals, Citrus and Deciduous Fruits, Lima Beans, Olive Oil, Honey, Etc.

SOILS AND THEIR ADAPTATION TO THE GROWTH OF GRASSES, GRAINS AND FRUITS.

[By George W. Coffin.]

Perhaps no country has a greater diversity of soils within the same number of square miles, than can be found in Santa Barbara county. Commencing on the mountains we find a grayish sand with some lime from sea shells, vegetable matter and some loam or light yellowish clay. On this, growing naturally, are live oak trees, many shrubs, or small growth of oak, ceanothus, wild apple, redwood, wild currant and gooseberry and several other varieties. There are plants too, such as the wild sunflower, morning glory, phacelias, thistles, ferns, etc.; and grasses, the cariso, or large bunch grass, several varieties of wood grass, aristada purpurea, or purple grass; and in depressions, the alfilleria and burr clover—the two most valuable forage plants known on the coast. Alfilleria is a Spanish word, and is generally pronounced "fillaree," accenting the last syllabie. It is a diminutive geranium of a delicious sweet flavor, eagerly eaten by all kinds of stock, and is not surpassed by the white clover of the East, in producing good-flavored milk, butter, beef and mutton. The burr clover bears sufficient resemblance to the red clover of the East to be readily recognized; its seed is borne in the axils or angles of its stems in a well-coiled burr. At maturity it falls, and the burr is sought all through the dry season by stock, and from the large amount of oil in the seed it possesses valuable fattening qualities.

In the slight depressions near the summits of the mountains, apples, pears, peaches, plums and grapes grow, producing good fruit. So little, however, has yet been done in the way of planting on these elevated lands that their capabilities are scarcely understood. At the present writing, at Pat's Station and at Marshall's, an altitude of about 3,000 feet, small orchards of these trees may be seen in a flourishing condition. There are other localities, some still higher, where the same fruits, as well as barley and wheat, may be successfully grown.

Following the mountain range westward, it terminates in grassy hills as it approaches the ocean above Point Concepcion. These hills are formed of a light-colored earth or clay, or of a dark, almost black mould, and adobe, and another soil more particularly described hereafter; giving out numerous springs and streams, yielding largely of alfilleria and burr clover, affording the finest stock ranges on the coast. The superior quality and abundant growth of these forage plants on the San Julian, Santa Rosa, Lompoc and other ranchos, give the region a reputation for choice beef, mutton, butter and cheese, second to none in the State. There is a firmness to this black soil that prevents waste by rains; yet in certain conditions of dryness it is light and easily worked. It absorbs water readily and retains it a long time. Similar soils cover the hills, slopes and valleys near Santa Barbara in many instances; and being protected by the mountain range (here from 3,000 to 4,000 feet high) from the north winds, and absorbing largely of the sun's rays, a difference of temperature occurs, favoring the production of many varieties of fruit which cannot be successfully grown in less favored places.

Before descending to the soils of the lower slopes and flat lands as they approach the coast, there is a soil so peculiar and so valuable that it should have a more careful examination and more particular description. It is called "diatomaceous soil," and in its greatest purity is limited in extent. Yet though the limits are narrow in which it thus occurs, it mingles with much of the hill and slope soil, affecting it favorably in all instances. Diatoms, of which it is composed, were formerly supposed to be the shells of a minute sea-fish, similar to the rhizopods that form the chalk hills of England. But recent careful investigation by scientists proves them to be of vegetable origin. They are the seed pods of a marine plant, that were produced in such quantities that they now form these narrow belts of greyish white earth; sometimes found in highest hills, or on slopes, and in places in valleys where the strata cross, to appear again in the hills on the further side; as it runs in an east-and-west direction parallel with the coast. It is found on both slopes of the Coast Range, from the Ojai to Point Concepcion and Point Arguello; also in the range on the opposite side of the Santa Ynez it can be recognized in the grey sides of peaks that face the valley. Sometimes branches of these strata may be seen on neighboring hills; they may be always known by the light grey color of their exposed surfaces. They are a silicate of lime; when dry they may be broken down by the thumb and finger into a powder; when wet, they form a smooth, adhesive paste. In the localities of their greatest purity, they exist in all conditions of hardness, from that just named up to the genuine old-fashioned gun-flint, used on fire-arms in early times.

This earth, considered as a soil, has, in a soft state, a great power of retaining moisture. "It holds water like a sponge," has often been said of it. Giving it off gradually, the streams are supplied through the long dry seasons; and as it mingles more or less with all other soils, except the real blue adobe, they are kept moist and productive without irrigation. This earth, mingling with a free black mould, forms the natural home of the alfilleria; with heavier black soil, burr clover does equally well.

Wheat attains to great perfection in both of the above soils, as instanced by Dr. Finch, on the foothills near the city, where 60 bushels to the acre were produced without irrigation. Also, by J. W. Cooper, of the Santa Rosa rancho, in the Santa Ynez valley; on 20 acres of new land 1200 bushels were harvested—this also without irrigation; yet much was lost by reason of imperfect machinery in the reaping and threshing.

[Mr. Thos. H. Hicks has just brought in from his 600-acre field in the same valley a bunch of wheat heads, most of them eight inches in length, and barley heads, six inches long, and all full of the finest and plumpest grain.]

In the summer of 1882, Dr. J. B. Shaw raised on his ranch in Los Alamos valley, 103 bushels of barley to the acre on 100 acres together. These are given as evidence of the productive capacity of this soil under favorable circumstances.

Among fruits, none finds a more congenial home on this soil than the olive. Ellwood Cooper of "Ellwood" is now reaping rich harvests in proof of this. The most of his trees are on the kind of soil known as the black tenacious vegetable mould and diatomaceous earth. This may be known by occasional cream-colored or ashy, angular, small-sized rocks, lying about loose on the surface. A bunch of wild oats has been hanging in the writer's office for a week that measured nine feet high, grown on soil of this kind. It is also the choicest of wheat land. Apricots, plums, apples, peaches and grapes thrive vigorously and produce abundantly on it.

The soil selected by J. Alston Hayne, Jr., for his olive orchard, near the Santa Ynez Mission, is the wash from gravelly and diatomaceous hills, and has also a black vegetable mould, which in combination forms a rich and grateful food for the olive.

It will be seen by the great extent of country influenced by the presence of diatoms, more or less, and the large and valuable products of grass, grain and fruits, that it should be, as it is, held in high esteem by all.

The writer speaks of its extent, as he has personal knowledge of it and no farther. It may exist in other places than those named. It is to be hoped it does, for the benefit of possessors and "the rest of mankind."

Far up on the foothills, back of Santa Barbara and of the Montecito valley, and extending well down into the valley, is a light colored soil of loam, mould and sand, not tenacious and hard, but what is usually called a "kind" soil. This is the home of the orange—in so far as yet tried—the true "orange belt." From here, (and occasionally from the orchard of Col. Hollister, which is in a valley between the foothills,) come some of the best oranges ever brought to this market. Until experiments go further and show differently, this must stand as a proven fact.

Valley land is generally too rich, causing a rapid and coarse growth of wood, leaf and fruit. When the underlying soil is gravelly so that the water may leach away and not stand about the roots of trees, irrigation may be practiced to beautify and swell the fruit, but it is always at the expense of flavor.

In the belt above named, there is an admirable adjustment of quality of soil, drainage, temperature and altitude to insure the greatest perfection in the tree and fruit. Irrigation is only adopted where loamy portions incline to too much sand, and then only in times of greatest drought. One singular fact in the production of the orange here, is that, in repeated trials between seedlings and the best budded varieties, by fair and competent judges, the former have been invariably declared the better.

Lemons also grow well in this, though doing better in a heavier soil. They sometimes yield enormously. In one instance, where the trees were scattered over an acre and a quarter, that would have occupied an acre had they been properly placed, in the orchard of Col. Hollister, there were picked in one season 60,000 lemons. Averaging one cent apiece, as they did when sold, gave $600 as the product of one acre in a single season.

Limes also do well on soils a little stronger than that required by the orange. But none of the citrus fruits should be planted on soils that are not or cannot be well drained. Water souring about their roots will soon be detected by what is termed gum disease, which is an exudation of gum, first on the trunk of the tree near the ground, and afterwards on the branches and even on the leaves.

On the hills and hill sides, where the earth noted above predominates, our best grape lands are found; even though scattered over with boulders, the soil being good between, they grow and produce heavily. Many prefer this boulder land, as the rocks absorb the rays of the sun, giving them off again at night, thus keeping a temperature above that of the valleys or unobstructed plains. The difference between the fruit from these slopes and that of the valley, is, in amount of saccharine matter, largely in favor of the hills or slopes. Market grapes are grown more than for wine, although about 40,000 gallons of wine have been made during the last year. No better soil can be found for raisin grapes. Some are made into raisins of superior quality, although the industry is not by any means established.

At the mouths of some of the larger cañons where the streams come out to a plain, almost level, they have in the course of centuries, brought down a large amount of worn down rocks, clay sand, and decayed vegetable matter. Distributing this admixture over the plains for several miles, it has formed the beautiful country of the Carpinteria and that called La Patera. In these favored localities the Lima bean production has become profitable. These soils seem especially adapted to their growth. Twenty-five hundred pounds have been raised on a single acre and large fields have averaged a ton to the acre. Even smaller beans which thrive in less valuable soil, yield better in this. The value of Lima beans has frequently reached $100 per acre in a single season. [This is an excessively modest estimate.]

The peculiar adaptation of this soil for the growth of Lima beans was first recognized by Mr. Henry Fish, who made a contract with an eastern firm to furnish all the beans he could raise for a term of years, at 4 cents per pound. The result is, that Mr. Fish is in very comfortable circumstances.

This soil is also the home of valuable orchards of apples, apricots, almonds, peaches, nectarines and plums. It seems to be especially adapted to the prune, as demonstrated by Mr. Eugene J. Knapp, whose cured fruit is unequalled in this market, and which, by his management, is largely remunerative. He grows the large and small French prunes.

In some parts of these spreading plains are slight depressions where more moisture is retained and where the soil is deeper and of a darker color. Selecting one of these depressions, some twenty-four years ago, Mr. Russel Heath cleared away the willows and began to plant the English walnut. His trees grew so well that he continued planting each year, until he has now some 8,000 trees, most of which are bearing. His sales in a single season have reached 35 tons, amounting to the snug sum of $7,000.

Ellwood Cooper has about 7,000 trees, and Col. Hollister about 10,000; most of the latter having been planted recently. It may be said of this nut that it delights in our best soils, the rich flats and banks of streams being the most favorable to it.

Eight years after Col. Heath's experiment in Carpinteria, Mr. Jos. Sexton began to plant the walnut in Goleta, and Dr. Brinkerhoff in Santa Barbara. All these orchards have grown well and bid fair to continue. About 85 tons of the nuts were shipped from this port last fall.

One fruit, well worth mentioning, is produced in localities like the above; it is the strawberry. No better soil can be found for it. Mr. Hemingway, of the Cathedral Oaks, has produced in one year, from an acre of ground, ten tons of delicious fruit; and has also, on the same acre, orchard trees growing, now some four years old, set at regular distances in true orchard form. He irrigates freely in dry weather and has "fresh strawberries all the year round" literally.

The Shepard brothers, in Carpinteria, are doing what Mr. Hemingway does in Goleta, and from these two sources Santa Barbara draws her main supply of strawberries. The variety most in favor is the "Monarch of the West."

From this free mellow soil, we turn to the genuine old grey adobe. It is more staunch than the hills, yet capricious in the working. If not handled at precisely the right time, it locks itself up for the season, and woe to the man who attempts to break through its closed doors. It dries and shrinks and breaks, leaving great yawning chasms between, into which the seeds of grains and grasses fall, to sprout and grow when autumn rains descend and cause an expansion of the rock-like clods till all is smooth again. When wet, it is too adhesive and cannot be handled; but taken at exactly the right time it proves a kind and pleasant soil, breaking into fine particles and remaining moist and mellow through the dry season. A finely broken soil, three inches in depth, covering the ground all over, prevents hardening and cracking, and in this condition, no soil is more productive. In this, corn, barley and hay give large returns; apples, plums and pears do well, and in some instances, the English walnut. Apricots are productive in it.

The lighter sands, near the ocean, are utilized by planting with sweet potatoes, which do wonderfully well in them; also with peanuts, which yield abundantly and are but little trouble to cultivate.

LEMONS.

[By Dr. L. N. Dimmick.]

That portion of Santa Barbara south of the Santa Ynez Range of mountains, lies in the choicest lemon belt of Southern California. This district extends from Point Concepcion to the Mexican line and includes all the protected localities within the equalizing influence of the ocean. The lemon and the lime are tenderer than the orange and will flourish only in a mild and equable temperature. The late B. B. Redding asserted that "it is not safe to plant lemon trees where the thermometer occasionally falls to 25 degrees; lime trees lose their leaves at 30 degrees, and the young wood is killed at 28 degrees." The late cold winters have proven that lemon and lime orchards are not safe in the interior away from the ocean. Even in the most favorable seasons the lemons grown in the interior are not as good and as attractive as those grown near the seacoast. The fruit of the latter is round, of medium size, the rind is thin and smooth, and the pulp is juicy. In the interior the fruit is larger and elongated, the rind is roughened and thicker, and the pulp less juicy. South of the Santa Ynez mountains the temperature rarely falls below 30° except in the lowest valleys. This district will eventually be largely devoted to the production of the choicest varieties of the lemon and lime.

LIMA BEANS.

[By L. B. Hogue.]

Some thirteen years ago a farmer in the then sparsely populated and undeveloped Carpinteria Valley, planted a few Lima beans and succeeded in obtaining both a good yield and a paying price. Others followed his example till fifty or a hundred tons were produced, which seemed for a time to exceed the demand for the article. But the demand grew apace, and the farmers continued to supply it till the product has reached ten to fifteen hundred tons annually. The different stages of the business as preparation of soil, care of growing crop and harvesting has been reduced almost to a science by our farmers, and they are now as prosperous possibly as any other people on the coast.

The seed is put into the ground about the first of May, in drills, and after some attention in the early part of the season, the crop is left to grow and spread over the ground until fully matured in the fall. The bean fields give the valley a rich and green appearance all summer long, and are a source of pleasure as well as profit. The roads leading to the warehouses and wharf are mostly level and hard, so that the produce is hauled to shipping at a trifling expense. The beans all go east via San Francisco. Yet we faithfully hope that a railroad will tap this section ere long, giving us a direct outlet to the east over the Atlantic & Pacific. The yield per acre of the Lima bean has usually been about a ton, but has fallen below that the past two seasons on account of a scant rainfall. Prices fluctuate in a manner quite interesting to those most concerned, and, although the price for the last two crops has not been so high as in some former instances, it is gratifying to note that it has kept well above the lowest record. The apparent tendency of the market is to seek a medium and become more steady as the business grows in extent. I suppose that three cents per pound would not be far from the average for the last two crops. By the advent of recent labor-saving inventions the cost of harvesting has been materially reduced.

OLIVE CULTURE.

[From Ellwood Cooper's "Treatise on Olive Culture."]

The following paragraphs are translated from the French of

Bertile: "The touching story of the flight of the dove from Noah's ark, related in Genesis, proves the existence of the olive tree in the earliest period of the world's history.

It was a celebrated tree among the ancients. It held the first rank in their mythology; Minerva taught the Athenians how to prepare the fruit, and they had a most religious respect for it. The Romans used the wood not only for fuel, but on the altars of their gods. It was the emblem of peace. * * * The olive tree transported from Egypt to Attica, belongs to the jasmine family, with evergreen foliage, small blossoms in clusters, and having some likeness to the elder tree flowering in June. It can be propagated in many ways, but the best way is by planting the seeds, and it is one which is practiced least. Except in damp soils where its roots rot, the olive grows everywhere. It accustoms itself to both dry and wet climates. Clay and mud are indifferent to it. Its long life is proverbial. In return, it takes thirty years, a man's lifetime, before it reaches its full capacity for bearing fruit. Of this tree, one of the most valuable gifts of nature, there exist 16 or 17 species, all exotic. Its fruit is oval, fleshy, with a hard woody seed enclosing a kernel. The meat, fine and covered with a green skin before its maturity, softens and becomes a purplish black in ripening; it is then that they grind them in the mill, then put them in a press to extract the oil.

With some exceptions one may say that in the Mediterranean Basin, from the 35th to the 43d degree of latitude is surrounded with a belt of olive trees. It is from this region that all Europe receives its oil for table use and light. * * * Rich in azote, and with considerable nutritive qualities, olive oil possesses, in the first place, the power of assimilating with the human body. It is instrumental in assisting in many medicinal cures where the method is cutaneous. It being more liquid than animal fat, always used for that purpose, it is easier to absorb. The injured parts, protected from the air by oily substances or salves, heal more quickly. These unctions give besides, more suppleness and elasticity to the muscles. As it is not penetrated by the poisons in the atmosphere, it is used with success in counteracting the deleterious miasma around swampy districts. It ought to be greatly preferred for the hair to pomades, as it acts more quickly on the scalp. Taken daily, by the spoonful, it is an excellent laxative to the system, and not tiresome to the stomach. * * * Mechanics refuse seed oils because of their dryness, as they gum up machinery, instead of greasing it and keeping it clean. It is just as important that the machinery of the human body should rebel against such oils. We ought to be familiar with the methods of extracting oil from all oleaginous substances, being so necessary to different industries. But all the table oil should give the preference to that made from a tree that the Almighty saved from the destruction of the Deluge and a branch of which the dove carried to Noah as a sign of forgiveness."

We refer readers to Mr. Cooper's Treatise for a full description of the manner of propagating, pruning and caring for the olive, as well as the process of making the oil and pickles. We can here but give a few extracts. In regard to the propagation of the tree, Mr. Cooper says: The common and preferred method is to plant the cuttings taken from the growing trees of sound wood, from three-quarters of an inch in diameter, to one and a half, and from fourteen to sixteen inches long. These cuttings should be taken from the trees during the months of December and January, neatly trimmed, without bruising and carefully trenched in loose sandy soil; a shady place preferred. They should be planted in permanent sites from February 20th to March 20th, depending upon the season. The ground should be well prepared and sufficiently dry so that there is no mud and the weather warm. In Santa Barbara, near the coast, no irrigation is necessary; but very frequent stirring of the top soil with a hoe or iron rake for a considerable distance around the cuttings is necessary during the spring and summer. About three-fourths of all that are well planted will grow. My plan is to set them twenty feet apart each way, and place them in the ground butt end down, and at an angle of about forty-five degrees, the top to the north barely covered. Mark the place with a stake. By planting them obliquely, the bottom end will be from ten inches to one foot below the surface. In Europe the trees are planted from 27 to 33 feet apart. My reasons for closer planting will be given in a subsequent article.

All trees, as a rule, should be propagated from seeds. The roots are more symmetrical, the tree not so liable to be blown over, and the growth more healthful; but I have not been successful in germinating them, hence I recommend the cutting. If trees are propagated from seeds, budding or grafting is necessary. I have seen the statement that it was necessary that the seeds should pass through the stomachs of birds before they could be sprouted; also that by soaking in strong lye the sprouting would be secured. I have not seen the result of either experiment, and accept the statement with more or less distrust. I presume cuttings can be obtained from any of the Mission orchards in the southern counties. * * * The cuttings will throw up numerous shoots or sprouts, all of which should be left to grow the first year, any disturbance of the top affects the growth of the roots. It would be advisable, however, where there are two or more vigorous shoots of about the same size and height from the same cutting, to pinch the tops of all excepting the one to be left for the future tree, so as to throw more force and vigor into that one. * * * Trees growing from cuttings will produce fruit the fourth year, and sometimes, under the most favorable circumstances, will give a few berries the third year. It is the habit of the tree to overbear, and as a consequence will give but little fruit the year following a heavy crop. This statement is verified by the most reliable books published on the subject in the French, Italian and Spanish languages. There are, however, exceptions to this rule in California. Mr. Davis, who had charge of the San Diego Mission orchard in 1875, assured me that he had gathered from the same tree, two years in succession, over 150 gallons of berries. I have also observed that some trees in my orchards have borne well successive years. The fruit-bearing can be controlled by the pruning. The cultivator will not forget that the shoots or branches must be two years old before they will give fruit, hence, partial pruning every year will give partial crops. My oldest orchard was planted February 21st, 1872. At four years I gathered from some of the trees over two gallons of berries. In 1878 over thirty gallons each off a few of the best trees, the orchard then being only six years old. In 1879, the crop was not nearly so large. I had planted several thousand cuttings in the spring of 1873, but these trees did not give at six years, a result equal to the first planting. The present crop, (1880) is quite good; the oldest orchard now being eight years, and I think I do not overestimate, when I state that the yield of some of the best and fullest trees will be over forty gallons. Trees large enough to give this quantity of fruit, planted at a distance of twenty feet, will occupy nearly all the ground, and therefore give all the fruit that can be produced on one acre. * *

The newness and richness of our soil will probably give, the first fifty years, double the best results given in those countries where oil making has been the business for so many generations. Our climate is congenial to the habit of the tree; it blossoms from the 1st to the 10th of May, and the fruit forms from the 1st to the 10th of June. At this season we have our best weather, free from extremes of either cold or heat. Nowhere in the world are all the conditions so favorable to perfect fruit-bearing. * * *

The olive usually ripens in November. In some localities in Eastern countries during favorable years, the fruit picking for oil begins as early as October, and for pickling, in September. In Santa Barbara the crop of last year, (1880) as also that of 1878, was unusually late in ripening, not being ready to pick before the middle of January—a delay of fully two months—the cause no doubt owing to the extraordinary rain fall of these two years.

HONEY.

A very comprehensive little paper on the history and progress of the bee business in Santa Barbara, written by Geo. A. Temple, then a partner in the Queen City Apiary, appeared in the pamphlet published by Joseph J. Perkins in 1881. The same was submitted to Judge Hatch, now sole owner of a number of apiaries scattered through the county, and being pronounced correct in the main, is here reproduced by special permission of Mr. Perkins:

The vicissitudes of bee-keeping have been many and varied, it having passed through the ordeal of haps and mishaps usual to the early history of almost all business ventures, and it stands to-day on a substantial basis of success, and an assured prosperity. Having no data, however, while at my apiary to guide me in such an interesting research, I shall be obliged to confine myself to such matters as I apprehend will be of interest or instructive to any who may contemplate embarking in the business here. I will simply say in passing, that bees were first introduced into this county as early as 1860 or '61, by Mr. Miner, who imported some eight or ten swarms, which he readily disposed of for the modest sum of $50 per swarm. Mr. Miner is also credited with having built the first frame house in Santa Barbara—northwest corner of Montecito and State streets. To Mr. Jefferson Archer, however, belongs the credit of being the pioneer bee-man of Santa Barbara county. Coming here in December, 1873, and bringing with him about forty-five stands of bees, he was the first to engage exclusively, and on an extensive scale, in apiculture in this county. Since Mr. Archer demonstrated that this county was eminently adapted to honey-raising for profit, many have followed in his footsteps more or less extensively and with varying success, and this interesting and profitable industry has been gradually extended until, with the close of the season of 1880, there were about 4,000 stands of bees in the county, which had yielded a product of rather more than 128 tons—(256,000 lbs.)—of extracted honey.

That portion of Santa Barbara county adapted to profitable honey raising, compared with the extent of territory devoted to this industry in some other counties. is limited, but the quality of its honey is unsurpassed, and while great advancement, both in the amount of product and methods employed in the apiary have been made since apiculture became a recognized industry in the county, there is yet ample scope for much greater extension. From the easternmost border of the county to its extreme western limit, honey-producing plants abound in profusion, upon the scarred and rugged face of every mountain and throughout the length and breadth of nearly every cañon, while upon each plain and in every valley, the glad humming of the "busy little bee," eagerly searching for nectar in the myriads of flowers there in bloom, makes glad the heart of the apiarist. The mountain redwood, the black ball sage, sumac, grease wood, coffee berry, etc., each in its proper time, furnishes abundant honey during the "building up" or brood-rearing period in the spring, but it is chiefly from the bloom of the sage family—the A.

nivea (button sage) and the A. polystachya (white sage) that the bulk of the honey crop is secured. These bloom profusely, and yield large quantities of honey from about May 1st until August 1st, during which time the entire honey product of the county is harvested. To persons with limited capital, who are not averse to dwelling amid the solitude of the mountains, and who can bring energy, perseverance and good judgment to their asssistance in the undertaking, apiculture offers large inducements. The possibilities in apiculture are immense, while the average profit on the capital invested, considering the amount of labor and time necessary to be applied in securing a crop, is greater than in many of the more pretentious industries of the county. A profit of four hundred per cent. on the investment has been realized from an entire apiary, while even a larger interest from individual swarms is by no means uncommon.

I have no desire, however, to give the impression that apiculture is one of the greatest bonanzas in the land, but wish it to be well understood that a fair profit may reasonably be expected by any one applying intelligently the latest improved methods of manipulation. Bees may be purchased in the winter and spring at prices ranging from $2.50 to $6 and $8 per swarm, and material for hives may be purchased at the planing mill in Santa Barbara cut and ready to nail together for 75 cents each. With a fair to ordinary season a good swarm will yield from 150 to 250 pounds of extracted honey, worth last season about six cents per pound net in Santa Barbara. Besides the yield of honey a good swarm of bees will increase two swarms in a season, while instances are not rare of a swarm—with its earliest increase—producing as many as five or even ten swarms in a season, to the infinite delight of the amateur apiarist. The following statement will give a better idea of what may be accomplished in this business with a very small outlay of capital:

Apiary debtor to one swarm bees	$ 5 00	
To one hive (all made)	1 00	
		$ 6 00
Credit—		
By 150 pounds of honey at 6 cents net	$ 9 00	
One swarm bees	5 00	
		$14 00
A net profit, at a very reasonable estimate of		$ 8 00

Or 133⅓ per cent. on capital invested—a very fair profit.

But like all other California industries dependent upon the rains for success, apiculture is subject to occasional drawbacks. An insufficient rain-fall, though perhaps stimulating plants to put forth the usual amount of bloom, lessens the quality of honey or curtails it altogether, while a general drought affects bees as disastrously as it does sheep, cattle or any other stock. The latter, however, is fortunately of rather infrequent occurrence, thereby enabling an apiary to recuperate from its destructive influence. Much of the success of apiculture depends upon the method or system employed in marketing the product; but as that subject does not properly come within the scope of this article I will leave it and simply suggest that a closer acquaintance between producer and consumer will conduce to larger profits on apiarian investments; and more systematic and united efforts on the part of the apiarists of the county to properly grade and market their honey, will tend to unprecedented apicultural prosperity.

FRUITS OF ONE RANCHO.

Scarcely as a representative farm, but rather as showing of what the country is capable in its highest development, attention is directed to Col. Hollister's thrifty plantations at Glen Annie. The following facts are kindly furnished by the proprietor:

Of citrus fruits there are upon the rancho 1200 orange trees, 500 limes and 500 lemons. These are cultivated entirely without irrigation; and to that fact is attributed the the extra fineness and intensity of flavor of the oranges. These trees are mostly seedlings.

Of deciduous fruits, such as apples, cherries, plums, peaches, nectarines, apricots, etc., there about 2500 trees; 4000 English walnuts, 1000 olives, 10,000 almonds. Of the Japanese persimmon there are 200 trees. This is a most delicious fruit, the result of the evolution of ages through better cultivation.

In addition to the above there are figs, guavas, loquats, shaddock and many varieties of ornamental trees and shrubs. Among the small fruits cultivated are fine strawberries, blackberries and red and white raspberries.

In the vegetable garden, green peas can be gathered every day in the year, and most other varieties of vegetables could be the same if properly cultivated. Irish potatoes do exceptionally well in this locality; sweet potatoes are very prolific and grow to perfection.

Of the cereals, nearly all kinds do well; Indian corn, especially, often producing one hundred bushels to the acre. In fact, so far as experiments have been tried, almost all food-producing crops were found successful. These experiments which have been spread over a series of years, at a cost of thousands of dollars, have been of great benefit to smaller farmers, who could not for themselves have made such expensive trials. One costly experiment was the tea-plant. Col. Hollister sent to Japan in 1872 for twenty-five bushels of the seed; importing also two Japanese tea-growers to attend the plantation. They actually raised 50,000 plants, which seemed to do well, growing from four to eight inches in height. As long as the warm weather lasted they flourished; but when the nights became cold they gave up the battle. It is the Colonel's opinion that a moister country, where sultry nights obtain, would be the proper climate for the plant.

Coffee was tried, but not to so great an extent that it can be called a settled question. Col. Hollister says that he did not give it a fair trial, and is of the opinion that it would do tolerably half way up the mountain sides. The date palm, now fruiting, seems to promise a success. It is a beautiful ornamental tree, at least, and for that purpose will always be a favorite.

The Glen Annie property comprises about 4800 acres; upon which there are now grazing about six hundred head of neat cattle, eighty horses and mules, five hundred hogs, and a thousand sheep. The dairy consists of one hundred and fifty cows, giving for constant milking about one hundred.

All fruit not disposed of in a fresh state is dried and packed for market on the place. There are employed upon the farm regularly about thirty laborers, whose compensation ranges from twenty to thirty dollars per month, with board and lodging.

MISCELLANEOUS INDUSTRIES.

Sericulture is an industry which is especially adapted to the climatic and other conditions of the coast and of Santa Barbara county. Here the first experiments were made twenty-five years ago. The first silk flag ever made in the State of California was manufactured from the cocoons of Santa Barbara silk worms. The mulberry trees planted by Messrs. Packard and Goux still remain to prove the theory here advanced. There is at present a renewal of interest in the subject, owing to the action of the Government, and there is some reason to hope that a sericulture station for experiments may be established here.

Pampas plumes are among our most notable exports—thousands are shipped every season to Europe and the East. It is a paying business. An acre of ground has been known to produce plumes which sold for more than $1000.

Free stone is another item by which the country will some time be made wealthier, when its quality and value become more widely known.

Oil wells have not, so far, been developed to any alarming extent, although petroleum certainly exists in many places. At times companies have been formed and efforts made to bring the oil to the surface in paying quantities, and in the neighboring county of Ventura, the wells have been rendered a source of profit.

Mining of all kinds has never got much of a foothold within the county, the only paying mineral product so far being the asphaltum, which is shipped in large quantities.

CATTLE, HOGS, POULTRY.

THOROUGHBREDS, DAIRYING AND WOOL-GROWING IN SANTA BARBARA.

THE HORSE IN SANTA BARBARA.

[By A. W. Canfield.]

A reputation was long ago made for this county, as one of the most desirable places in the world for the breeding of horses. The native horse, foaled among the wild beasts of the mountains, came into the world to fear every sound; many a colt in those days fell a prey to the lions or grizzlies which roamed among the mountains. Those which survived gamboled among the oaks and pines, feeding upon the rich grasses on the rolling lands, in this dry and even climate—all these points making up the requisites which are sought for by those who have made a success of horse-breeding.

The staying powers of the native horse were wonderful. Only a few years ago an old "pinto," which had daily delivered groceries about town for many years, weighted with a California saddle of 50 pounds, besides his rider, made the distance between Santa Barbara and Los Angeles—over 100 miles—without shoes, between sun and sun in one day!

Another ran four mile heats in Los Angeles county in 7:30. The last quarter of the last mile was covered in 19¾ seconds. Score upon score of such instances might be noted where a California horse, raised in this county, has performed deeds of speed and endurance that would arouse enthusiasm anywhere in the world where ever the horse is known and appreciated. Seventy-five miles a day is no uncommon ride for a Santa Barbara horse and vaquero to make; and these same horses are turned loose in a corral or picketed upon the native grasses, without further care, the next day to buckle again to his usual work.

The "iron age" for the native horse came with the introduction of staging through the country; no farrier ever crippled the nimble feet of our mountain steeds, until the advent of civilization with all its cunning artifice, which came to war upon slumbering nature. In quality, shape and limberness, the foot of the horse raised upon the inland ranges of Santa Barbara county, is surpassed by no class of horses on earth. Since we have supplanted the native Californian with the so-called "American horse" the quality of bone and of the foot is quite noticeable; for the usual splints are few, and quarter cracks and contraction and ails are almost unheard of.

The American horse matures a year earlier than in colder countries; two-year-olds are equally developed with three-year-olds in New York. The writer has broken to harness many two-year-olds upon his farm, when they went to work alongside of six-year-olds, doing their full share without any appearance of suffering.

It is, really, not our breed of horses, but the breed of climate, grasses, soil and water that makes the horse of the country. In this enterprise of horse-breeding Santa Barbara still offers inducements to capitalists. The markets of the great southwest and Mexico, with our own rapidly settling country, will absorb every fine horse that can be raised for many years to come.

HOGS.

[By L. Babcock.]

The following practical suggestions on hog-raising are extracted from an article on the subject, prepared by Mr. L. Babcock for the INDEPENDENT:

The question is frequently asked me, is the raising of hogs a good business in California. I answer, yes. My reasons are as follows: Some would say that if everybody went into the business the supply would be too great, and there would not be a sufficient demand. But that is impossible, for two reasons. First, the great grain-producing valleys of California are nearly destitute of running water; second, the cost of fencing is too great. The laws of California protect a man's crops, and but few farmers will go to the expense of fencing.

Again, the world has had no devastating wars or fatal diseases for the past nineteen years, and with the rapid increase of human souls, our beef and mutton supplies are not in fair proportion to the requirements of the meat-eating races. Then our large cattle and sheep ranches are being subdivided and sold out to small farmers. We all know, or least I do, that extensive cattle-raising is almost at an end in North America. I will state here that in 1841 nearly all of the country west of the State of Indiana was one vast expanse of prairie, grazing lands, and parts of it were covered with an unlimited number of buffalo, elk, deer, antelope, etc., that furnished the aborigines or Indians with their supplies of food for the sustenance of life, and now we have to feed them (the Indians) on beef.

The reason why I referred back to 1841 was to show the young reader and others how there has been a steady increase in the demand from that time till this, for meat and grease; that prices have kept pace with the demand; that the increase of our mining industries make an increased demand for bacon and lard, bacon being the only meat convienient for the miner. As he digs metal out of the earth, he does not grumble at the price of his daily bacon. Our desert and mining lands must depend upon farming localities for their food supplies, as we upon them for the metals.

Hogs can be raised here with little trouble after you are prepared, as we do not have any or but few storms during each year, and no fatal diseases such as cholera. Neither have we any trichinæ in the bacons on this coast.

On the 19th of May, 1881, I purchased 120 acres of land in the Lompoc valley, all fenced and improved ready to go into the business of raising and preparing hogs for the market. I also bought 600 head of hogs, big and little, and the growing crop, at a cost of $13,066. I raised grain on 100 acres of the ranch. On the last of August, 1881, sold to Sherman & Ealand, of Santa Barbara, 302 head of hogs. They received them on the ranch and paid me $1962.50. In September, 1882, I shipped to San Francisco, 323 head of hogs, off the same ranch, and sold them for $3801,26, and after deducting all expenses of driving, shipping, commission, etc., I got a net return of $3,284.63. And I have 100 or more still left on the ranch.

The past three years have not been extra for grain raising, but fair. You can figure the profits for yourself. Poor or medium lands would not give so good an income, but there would not be much capital invested as the land would not cost so much per acre, though the fencing would be the same. A band of hogs, large or small, are not destructive to a grain field when turned into it at the proper time, or to stacks of reaped or headed grain that has been prepared for their use. The hog wastes nothing—except young chickens, ducks, turkeys and lambs, and sometimes "cultivates" the flower garden. He is a great expert in "legerdemain." Give him all he wants to eat and he is quiet, and soon ready for market. The quicker you get your money out of him, the better. Never hold on for an extreme high price. When the hog is fat, sell him, and go for more. In conclusion, let me say that you must have land, water and good fences, to make hog-raising profitable.

DAIRYING.

[From the INDEPENDENT of Dec. 15, 1884.]

During the winter season, while grass is plenty, is the busy season for the north county dairymen, and the amount of butter produced is then the highest. San Luis Obispo county has for some time been the headquarters for dairy interests of Southern California, and it is only of late that Santa Barbara county has claimed any importance in this branch of industry. But the Swiss of San Luis, as also the Americans in the business, have been gradually spreading out and down until there are dairies down as far as Las Cruces. In conversation with Antonio Righetti, proprietor of the Najoqui dairy, near Las Cruces, a few facts were learned in regard to the profits of butter making. Taking one dairy as an illustration, of 1,500 acres, upon which 150 cows are milked during the winter, and fifty or sixty during the summer, it will be seen that there is much money to be made, if properly managed. During the season of ample feed, the amount of butter produced upon the Najoqui is about 185 pounds per day. This is shipped regularly once a week to San Francisco by steamer from Gaviota in boxes, and nets in San Francisco to the dairyman, 21 cents. This makes a return of $38.85 per day for butter alone. Besides this, the proprietor feeds a calf for each cow, which becomes a source of profit, in the neighborhood of $10 each per year, or about $1,500 per annum for the calves. Besides this, about May, after the calves have been turned from milk to pasture, there is sour milk enough to fatten eighty hogs. These are worth five cents per pound. A hog weighing 100 pounds when first bought, in a month should weigh 200 upon the rich healthful diet, making a gain of 100 pounds at 5 cents equal to $5 on each hog, or $400 on the lot as a month's profit. In these three items, butter, calves and hogs, the profit lies. The income from them depends upon the skill, energy, intelligence and industry of the proprietor. The Swiss, judging by their past success and present growth, are gaining the ascendancy, probably from more thrifty habits. Upon the ranch taken as an example $4,000 has been cleared in a year. The capital invested in land, cattle and fixtures was $2,500, being a profit of 17 per cent. But this was not all. At the close of the year the property had almost doubled in value through improvements in the land and stock and increase in calves. Only six men employed at $30 per month are required upon the Najoqui and eight horses for saddle and four work animals. These figures are given in a rough shape as it is impossible to calculate closely upon a business where the individuality of the proprietor makes such a difference. But it is a significant fact that after ten years work one Swiss dairyman has rented his ranch and returned to Switzerland, living upon the interest of his money. The business is not difficult to learn. We are told that none of the Swiss who now do such a large proportion of the business understood it when they first came here, but learned it simply by working upon dairy ranches. The industry now looks more promising than ever. Butter being always in demand makes an unfailing market. The soil if properly managed may yet solve the summer problem and an equal number of cattle be milked all the year round.

POULTRY.

[By A. W. Canfield.]

We will offer an occupation to every man and woman able to do anything. Sometime the question may come to you, What shall I do when I get to Santa Barbara? Let me answer! You can raise poultry with profit near our growing city, with pleasure and profit. Fowls are in ready demand at our hotels and restaurants the year round. It is now May 5th, and chickens or fowls are bought at $7 per dozen; eggs are always in good demand, the highest price of the year being fifty cents per dozen, the lowest price fifteen cents. The average price may fairly be estimated at thirty cents per dozen. With these very fair prices you calculate that a brood of chickens may be raised from chickenhood to maturity through any part of the year in the open air with great success. Mortality in chicks about Santa Barbara is very slight when proper care and feed is given the fowl and brood. Cleanliness is the main thing the denizens of our poultry yards demand of their keepers. Everything else will go well if the proprietor of "commercial poultry keeping" will observe this law. Grain, vegetables, butchers' scraps, hotel and restaurant refuse can always be bought at a price that pays to feed fowls about Santa Barbara.

The incubator is receiving considerable attention from those so enthusiastic that numbers and labors and care are of no manner of account until the problem is thrust squarely before them. Our matchless climate, so even after sun down to sun rise, allows the incubator to do the best possible work. The hatch from the incubators the past season is fully equal to the same work done by the hen-mothers. The county is stocked with a fair quality of common fowls and the yards of our local fanciers afford variety equal to any, and excellence of type that will satisfy the most exacting judge of high-grade fowls.

For the delicate man of moderate means no better nor pleasanter occupation offers itself than the keeping of a number of fowls. Any day that one is well enough to be about, the care of poultry is not detrimental to the most sensitive nature. Occupation that employs the hands takes care off the mind and relieves that depression that always goes with ill health especially when a person has no employment. Cheap lands and snug little homes can be had near Santa Barbara suitable for poultry keeping. Those who have kept fowls elsewhere and will try them in Santa Barbara will have no trouble in starting in the right direction. To the novice we would offer a word of caution, "go slow;" observe the differences in the changes of country, and allow the poultry business to "grow to you," and it will be found another among the long list of useful pursuits which is preëminently adapted to Santa Barbara county.

HUNTING AND FISHING.

SPORT ON LAND AND SEA IN AND ABOUT SANTA BARBARA.

GAME.

Santa Barbara is preëminently a game county. In the proper season wild geese and ducks, snipe, quail and rabbits (both cotton-tail and jack rabbits) are plentiful, affording both pleasure and profit to the scientific Nimrod. Along the Santa Ynez river and among the foothills deer are likewise to be found in numbers; and twenty miles from the city the grizzly may be interviewed by the cool and cautious hunter who desires such an interview. There, too, and up in the Cuyama country, may be found the mountain sheep, the coon, the wildcat, the coyote and the California lion, which is really a cougar. The gamy trout abounds in mountain streams, and all these at a short distance from civilization, easily reached on horseback, the best hunting grounds not being favorable for the passage of wheeled vehicles.

THE FISHERIES.

The Santa Barbara Channel, which includes the coast from Point Concepcion to Ventura, bordered by the islands of San Miguel, Santa Rosa, Santa Cruz and Anacapa, as a fishing ground has no superior on the whole coast of California. The islands are fringed with kelp; so also is the main shore, and this marine growth gives security and feeding grounds for the rock cod, red fish, craw fish, smelt, flounder, halibut, sardine and other varieties. In winter mackerel of fine flavor are taken; during the months of June and July the delicate pompano is found in abundance, and during the season (summer) the barracuda, bonita and albacore are taken by trolling in sail boats. It requires a stiff breeze to decoy them to the pretending sardine, and pleasure combines with profit in a day so spent upon the blue waters of the channel. Parties are formed almost daily during the season, who are taken out by one of the pleasure boats, schooners and other craft, to spend the day upon the water, coming back at night with loads of fish and sunburnt faces.

The abalone trade is worth mentioning as a separate industry. The shells are gathered upon the islands and shipped east to be wrought into jewelry. The clams of the coast have been noted hitherto; there are also mussels in quantity and traditions of oyster beds about the islands.

The seal and otter hunters, for whom Santa Barbara was once noted, still ply their adventurous trade; but the game they seek is no longer plenty in the old haunts about the islands. The otter in fact have about disappeared from these waters; but the seal rookeries on the different islands are yet of sufficient importance to tempt an occasional crew of hardy mariners.

As a sea-port town, Santa Barbara has naturally a number of craft always lying near the wharf or flying here and there about the channel, and a favorite pleasure trip is that of crossing the channel to the islands, of which much might be said. But we have already spun out our descriptions to a far greater degree than was anticipated and circumstances over which we have no control oblige us to come to a stop.

Wool Growing.

It is not many years ago that this county was preëminently a sheep district. From paragraphs scattered here and there throughout this pamphlet, it will be gathered by the sapient reader that wool-growing no longer occupies the ranch owner to the exclusion of everything else. The industry was introduced by Col. Hollister, more than twenty-five years ago, he bringing a band of sheep across the continent as his sole wealth. The story of his experience has been often told—a story which must interest every one, on account of the dogged perseverance displayed, with which he clung to the business, against all manner of drawbacks, until he became the owner of thousands of acres of the finest land on earth, and these acres inhabited by flocks and herds of almost fabulous value.

But the business of sheep and wool growing is, as we have before intimated, no longer of the first importance. It requires too much land to be entered into without considerable capital, and those of small means will scarcely think of it. We had contemplated a full account of the industry from its first inception, but have been obliged reluctantly to abandon the idea.

There are now, in fact, but two or three large sheep-owners remaining in this county—the Dibblees, J. W. Cooper and Col. Hollister being the only ones owning them by the thousand. This, however, does not include the islands. Both Santa Rosa and Santa Cruz are largely stocked with sheep. Small schooners owned by the proprietors are kept busy between the islands and the main land, and steamers also touch at the island wharves to carry the wool and mutton to the San Francisco market.

AN ADDRESS

Delivered by Hon. Chas. Fernald, Mayor of Santa Barbara,

Before the Agricultural and Horticultural Societies of Santa Barbara, at Lobero's Theatre, October 11th, 1883.

Mr. President, Ladies and Gentlemen of the Santa Barbara Agricultural and Horticultural Societies: I believe I ought first to congratulate you upon the success of your fair. This display of the varied and rich products of our home farms and orchards is most satisfactory and encouraging. The bounteous profusion which we see around us must fill our hearts with gratitude to the "Lord of the Harvest," and strengthen our attachment to this greatly favored land. Nor can it fail to increase our interest in the development of our agricultural, horticultural and pastoral resources.

Great changes have, indeed, taken place in the employment and pursuits of the people of this county during my residence here. This rare exhibition is, perhaps, for that reason, more interesting to me in many ways than to most of you who belong to the period of the renaissance, so to speak; while I may be classed with the pioneers, a very select few who came a little in advance to gain the pleasure of welcoming you on your arrival to this land of fulfillment. We knew all the time that you would follow, although some of us were well-nigh out of patience at your delay.

The principal pursuit of the people of this county, as well as of all this part of our State from the year 1852 till 1863 (as it had been long before), was the raising of neat cattle, horses and sheep. At all, or nearly all of the Missions there were flourishing orchards and vineyards, with acqueducts and ample water supply for irrigation. There were a few vineyards and some orchards in sheltered places here and there on some of the great ranchos, as these estates were called, where some good wine was made and olives of excellent quality were preserved. Almost every rancho had also its fenced field where a little wheat, barley, corn, squashes and melons were raised for home use.

In the town here, from the time I came till about 1869 or 1870, the vegetables consumed, for the most part, came to us by steamer once a week from San Francisco.

The business of raising cattle in large herds in Southern California may be said to have come to an end by the great drouth in the year 1864. Nearly all of the live stock perished in that year. Land without cattle was worthless, and many of the principal land owners were impoverished.

Farmers, what few came in those early times, met with obstacles and difficulties which no longer exist. You can hardly realize how great they were. In a country over run by wild cattle and horses, no farming could be done without fences to protect the growing crops. Fencing was very expensive by reason of scarcity of material and cost of transportation. And then if more of a crop was raised than was wanted to supply the home demand, a loss was suffered by the experimental farmer; for the expense and difficulties of transportation cut off competition in other markets. And again, if the crop were grain, as soon as ripe in the field, a controversy would arise between the farmer and the squirrels to determine which party could get possession of the most of it. In these contests the farmer generally got the worst of it, for he would spend about as much money and time in efforts to destroy the vermin, as his share of the crop was worth. Profitable crops and fruits could only be determined by continual and expensive experiments in this new land.

Despite all these obstacles and difficulties it may be said that from about the year 1870 the farming and fruit interests began to develope and to merit recognition as a source of wealth; and notwithstanding some failures and discouragements, have continued to increase steadily till the present time. We see around us to-day the most assuring evidence of permanent success.

The farming interest in this county now predominates over all others.

I am enabled by the painstaking labor of a friend, to present a statement of farm products for the present year, based upon estimates of the year 1881, which may be accepted as approximately correct:

Wheat, bushels,	950,000;	acreage,	40,000	Beans, bushels,	100,000;	acreage,	8,500
Barley, "	675,000;	"	25,000	Potatoes, tons......	2,000;	"	450
Oats, "	1,200;	"	75	Sweet " "	150;	"	20
Corn, "	95,000;	"	25,000	Onions, bushels...	175;	"	¾
Buckwheat, "	1,200;	"	40	Hay, tons.............	2,000;	"	5,500
Peas, "	200;	"	5	Flax, pounds......	275,000;	"	300

Our grain producing district lies in the northern part of the county, embracing the great valleys of Santa Ynez, Los Alamos, Lompoc, Santa Maria and Guadalupe.

The wool, cattle and dairy interests stand next in rank in point of value, the sheep numbering about one hundred and fifty thousand, and the neat cattle not falling short of twenty thousand head, mostly in small bands and of approved breeds.

Fruit and nut bearing trees may be fairly estimated to cover three thousand acres. Our rich and sheltered valleys afford ample space for fruit culture. The cannery and the fruit dryer assure large profits to the producer by reducing bulk and cost of transportation to other markets.

The cultivation of the walnut promises large returns to those who can afford to wait the growth and maturity of this valuable tree. In the very fertile and deep soil of the valley of "La Carpinteria," at "La Goleta," the "Dos Pueblos" and many parts of Montecito this nut tree, the olive, and many other fruit trees yield abundantly, and of the best quality, without irrigation. Indeed those valleys produce a greater variety of excellent fruits than any locality with which I am acquainted; due, perhaps in part, to the adjacent protecting range of mountains. These places being almost exempt from frosts, many varieties of trees, the olive and citrus especially, are liable to be injured by insects and fruit pests of other and warmer climes. This impending danger must be warded off by the combined action and honest work of all fruit raisers; for all are equally interested. Some provision ought to be made for condemning and destroying, if need be, a neglected and hopelessly infected orchard, for the common good, as all other nuisances are dealt with. The insect-devouring birds should be protected at all hazards, as they are the most constant and serviceable auxiliaries of the fruit raiser.

I must claim the privilege of suggesting to those who are fortunate enough to possess a few acres of surplus land, the policy of planting out in angles and corners, and along the borders of your farms valuable timber trees, such for instance as the black walnut and hickory, which appear to thrive in our deep soils as well as in the lands east of the Mississippi. The hillsides, ravines and parcels of what seem waste land afford room for many valuable varieties of pine and other forest trees, giving beauty and value alike to the farm.

With the coming of good farmers and fruit raisers, of course there came into existence these twin societies, so necessary to promote these great industries in our own county, and by comparison and competition to advance the same interests elsewhere in our State. In this way the experience of one farmer or fruit raiser in the choice of land, methods of cultivation, selection of crops and of fruit trees, comes to be multiplied indefinitely. And it seems to me that the collection of authentic statistics by your societies, of the products of our farms and orchards and vineyards, would prove of very great value.

These annual competitive expositions create pride in and respect for the farm and for its varied employments. Agriculture and horticulture do not admit of monopoly; and the competition of the farmer is about the only one that does not engender strife or bitterness. Success of one does not imply defeat of some other competitor for the best products of the field. In other vocations men try to out-wit and get ahead of each other. The farmer must of course exercise good judgment in buying his land. Soils must be examined, exposures considered, means of transportation, and sanitary questions not overlooked. Poor land means a life of labor and a constant struggle with poverty.

Quality rather than quantity of crop makes the successful farmer, in the long run. All must see how far these competitive expositions go to prove this proposition. All now look for the best of everything. The work of producing and harvesting a crop of bad, is as expensive as one of good quality; and a prudent and wise farmer will content himself with being a producer, selling his crops seasonably, and never tempted to speculate in his own products, nor to compete with the grain merchants with whom he must always be at a disadvantage. He might as well try to control the rainfall as to attempt to fix the price of commodities in the great markets of the world.

In this part of our State where the annual rainfall is often scanty, a stream of pure water, or some copious water supply is an element of great value to the farm and to the orchard. A field of alfalfa or millet is a mine of wealth to the farmer in a rainless year. By it he is enabled to keep in good condition farm stock that would perish without it. No farmer should be without a few head of very choice stock. It is the truest economy. The cost of raising a fine animal is no greater than an inferior one. The exhibit of well-bred stock here, shows that you appreciate this important fact.

The manufacturing interests are great; the combined transportation enterprises by land and sea are great; but the farming interests are yet greater than all, for all rest upon them. With what feverish anxiety does the world await the first reports from the harvest fields in this country. A few wet days, more or less, before harvest in England affect the price of grain all over the world. For if the crop is short there, it is known that English grain merchants must be in the market to buy enough to make up the deficiency. Indeed, of late years, England has relied upon us for her yearly supply of grain. Little did they think when the "May Flower" set her sails, that the history of Joseph and his brethren might again be repeated and gain a national significance. Much of our wheat finds its way to France and Germany also. Our agricultural resources excite the greatest interest everywhere. As an evidence of this we have seen that in 1879 the British government sent a commission to this country to examine the grain producing area, extent of supply and methods of production.

The work has been accomplished and the report of the commission is soon to be made public. It cannot fail to be as advantageous to us as to our customers across the sea. While with us it is a question what to do with our surplus, in England, and not infrequently in France and Germany, it is a more serious matter where to look for a supply. England and Germany have not land enough for the people. In England there is coal and iron in abundance, and multitudes of skilled workmen who must work day in and day out for bread. They can build ships cheaper than any other people, and have well nigh monopolized the carrying trade. The largest part of our surplus grain is taken yearly in their ships to foreign markets.

We see that the British Isles, with their thirty-three millions of people, have but two and one-third acres of land to each inhabitant. It is not surprising that the ownership of land in that country should give social rank and influence. How best to cultivate these few acres, (exclusive of the great parks and forests of the noblemen) to furnish the largest supply of bread and meat from year to year, is the vital question there. Hence agriculture is studied by some of the ablest men. It is there a question of chemistry, engineering and drainage, as well as of economy and finance. It is not of so much importance what the present crop is to be (as with us,) but the next and for future years. Our soils are rich and we do not heed the future. There they must restore annually—pay back—something to make good the waste and restore the equilibrium of productive power. Some years ago it was stated upon undoubted authority, that the value of the animal manure yearly applied to the crops in England, at current prices, surpassed the whole amount of its foreign commerce. The land in that country is in the hands of a few persons, hence the rates of rent assume very much the same importance that the transportation question does in this country.

The whole land of France is suffcient only to give its thirty-seven millions of people three and a half acres; while Germany, with forty-three millions, has but three acres to each person. All these countries do not make up the area of the Mississippi valley, and their total populations might be thrust into that valley and Texas, and yet there would be room for more. The people cannot buy land in those countries and hence they come to our new territories where lands can be had at low prices. And there can be no doubt that immigration to this country is favored by the English and German governments, as a temporary means of postponing and escaping the difficulties of the labor and land questions. The titled classes, controlling the State with a standing army, will not surrender to husbandry their parks and manors.

Our population in 1880, was a little more than fifty millions. We had land enough then to afford fifty-eight acres to each inhabitant.

Compare the aggregate wealth of those great nations with our own and the result is very much in our favor. In our country the total amount reaches forty-nine thousand seven hundred and seventy millions of dollars.

Great Britain comes next with forty thousand millions. France and Germany follow, each with about ten per cent. less.

In the United States there would be $995 to each person. In Great Britain $1,160 to each.

The remuneration of labor in all these countries cannot be without interest to us. In this country, assuming the product of labor to be 100, seventy-two parts go to the laborer, twenty-three to capital and five to the Government in various forms.

In Great Britain, fifty-six parts go to the laborer, twenty-one to capital and twenty-three to the Government.

In France, forty-one to the laborer, thirty-six to capital and twenty-three to the Government. In Germany the result is nearly the same. Sixteen per cent. more of the produce of labor goes to the laborer in this country than in England, and thirty-one more than in France and Germany.

Everything that relates to the public lands, the common inheritance of all, and its disposition by Government, must be of great interest to farmers. And all must see that our national legislation, tending to create monopolies in land, is a serious blunder. The grants by Congress to railroad corporations within the last twenty years, more than equal in area all of the New England States. And under the so-called "timber act" passed ostensibly to promote the growth of timber, nearly all of the accessible and valuable timber lands of our northwest coast have passed into the possession of a few mill owners by a scandalous abuse and perversion of the law, bad as it undoubtedly is.

This fine timber, the growth of ages, will shortly disappear by the steam saw mill, by waste and by fires.

In the nature of things every wrong and error, individual and national, exact a penalty. As our population increases the demand for land will increase and it is certain we shall repent of the folly, although to avert the consequences may not be so easy. It is true that our laws of succession tend to favor the division of estates, offering a slow but peaceable remedy for the accumulation of land in individual cases, but they do not affect land-holding by corporations under congressional grants.

It may be said that it is the purpose of the corporations to sell these lands; but it rarely happens that patents are applied for till settlement and improvement of adjoining sections sold by the government to farmers have created a demand for them, which the corporations take advantage of in fixing prices. And it is understood that no taxes are paid on these vast grants listed to the corporations till patents are issued by the Interior Department. These grants were and are feudalistic in tendency and productive of evils which the wisest English statesmen are seeking to abate in Great Britain without a revolution.

Our government is founded upon the theory of the greatest good of the greatest number, or to speak more accurately, the greatest good of all. We have seen what has been accomplished by it—obstructed as it has been in some of our great cities by what may not be inaptly called, untimely crowding from below. The remedy for this evil is to broaden the republic at the base by increasing the number of farmers who own the lands they cultivate and not by any system of agricultural serfdom, which is but a substitute for slavery.

The owners of the lands of a country will ultimately shape its legislation and control its policy and destiny, and it is one of the hopeful signs of the present day to see so many educated business men, and especially young men, buying lands to become farmers and fruit raisers. Tenure must be fixed and not subject to change.

Ownership of land increases industry and desire to improve it. I suppose there never was a tenant who did not desire, in his heart, to acquire his landlord's title in some way. Ownership of lands in reasonably small parcels, by the farmers who have worked their acres is the secret of the marvelous growth of New York, Ohio, Illinois and the northwestern States.

"Slavery," it is said, "has its lash," but no inducement to labor other than the ownership of property, has yet been devised by the wit of man. In commercial pursuits but few attain permanent success, even at the cost of life-long efforts. A large percentage fail outright. In manufacturing, laborers must be more or less a dependent class. Peace or war, the state of trade and commerce, public disaster, high or low tariff, business panics and the like, affect employment, and make them necessarily dependent upon others for the means of support. And in cases of failure of employment the suffering and want will be great in proportion to the numbers affected. The farmer, on the other hand, made secure by his own labor—the highest form of title—in the possession of the product of his own acres, must be regarded as the most fortunate and independent.

In all ages there have been theorists and dreamers who have planned to put an end to poverty. But the farmer alone is the possessor of the secret. His own instructed labor, with the rains, the dews and the sunshine, have given it into his possession and committed it to his keeping as long as he will. He has yet more. As a compensation for his toil he has the infinite and pure pleasure of witnessing the mysterious birth and development of plants and trees in all their varied and beauteous forms. "The wild roses of the wood" breathe upon him, the apple blossoms and the new-mown hay give him their matchless perfume, and the ripe fruits in their season furnish forth his table.

AT THE FAIR.

Among the products of the soil,
The fruits of free and honest toil,
St. Barbara's wheat and wine and oil,
  What shall a landless poet bring,
Who has but heaps of heathen spoil
    For harvesting?

The artist's dream of woodland grace
Or memory of a lovely face;
The maiden's web of filmy lace;
  The housewife's skill, the mysteries
Of modern art, are all in place
    And safe to please.

The flower and fruit of every zone
St. Barbara welcomes as her own.
And her enthusiast sons have shown
  A courage in them lurk
For greater victories to be won
    Through honest work.

But we, who have no tenancies
In golden grain or olive trees;
No house nor land; instead of these,
  In airy towers of fairy gold,
We weave but phantom tapestries
    Against the cold.

O then, while farmers bring the spoil
They gather from a fertile soil,
St. Barbara's wheat and wine and oil;
  A landless poet can but bring
A simple rhyme, not worth the toil
    Of harvesting!

MARY C. F. HALL-WOOD.

Nov. 9, 1881.

Record of Temperature at Santa Barbara, Cal.

HIGHEST.	1878	67	69	69	77	77	76	80	84	94	84	76	76	94
	1879	68	79	89	83	92	97	80	85	86	93	77	73	97
	1880	70	69	68	69	82	75	75	77	85	81	77	70	85
	1881	66	78	76	76	74	78	87	84	80	73	73	72	87
	1882	62	68	78	72	78	77	80	81	83	82	75	80	83
	1883	72	77	72	82	92	92	79	84	98	75	80	70	98
LOWEST.	1871	40	40	46	46	52	56	58	60	54	48	40	40	40
	1872	38	42	46	47	50	57	60	60	58	49	51	41	38
	1873	42	40	48	46	54	58	58	61	56	50	46	40	40
	1874	38	41	41	50	53	56	58	58	57	52	48	43	38
	1875	40	41	44	41	50	54	68	69	57	52	48	43	40
	1876	40	41	46	47	53	56	59	60	56	55	46	43	40
	1877	42	46	46	50	52	58	61	58	58	46	49	40	40
	1878	38	41	46	46	52	57	58	57	56	52	41	37	37
	1879	38	39	46	50	52	54	58	59	55	50	42	33	33
	1880	33	36	39	46	51	52	53	54	51	41	36	41	33
	1881	40	43	43	46	52	55	57	55	53	45	36	39	36
	1882	33	33	42	43	50	53	52	56	50	47	38	38	33
	1883	33	31	47	44	50	52	56	55	58	47	41	40	31

Average for January, 53.25. Average for July, 68.15. Yearly Average, 61.43. Difference between July and January, 15 degrees.

Dec..	4.26	.57	1.41	6.56	4.34	5.26	....	.31	....	3.65	0.89	5.01	0.04	.96	.10	2.89	3.20
Jan..	3.26	.25	.86	2.53	.58	4.54	14.84	7.56	3.04	7.87	4.83	1.42	3.02	1.30	2.18	6.54	4.06
Feb..	2.12	5.87	2.92	1.81	5.48	3.17	.18	5.67	....	12.32	.72	11.50	.30	2.85	2.95	10.97	4.30
Mar..	4.22	.83	.02	.18	.05	.78	.38	2.73	.61	2.68	.34	1.22	1.36	6.85	3.27	8.92	2.15
April	.46	.99	2.02	1.80	....	.28	.10	.27	.39	3.34	1.80	6.25	.63	2.82	3.3	2.02	1.51
May.	.20	.74	.37	....	....	.14	....	....	.45	.29	.30	....	....	....	2.65	.49	.35
June	....	.07	....	.14	....	....	....	....	....	.05	.11	....	....	....	.35	1.75	.15
July	....	....	....	....	....	....	....	....	....	....	....	.07	....	.00	....	....	.01
Aug.	....	....	....	....	....	....	....	....	....	....	....	....	....	....	....	....	....
Sept.	....	....	....	....	....	....	....	....	....	....	....	.11	....	....	....	....	.01
Total	15.77	10.27	8.91	14.94	10.45	14.44	18.71	23.07	4.49	31.51	15.34	27.95	15.68	16.93	12.97	35.53	10.32

Average annual rainfall for fifteen years, 17.31 inches.

Comparative Annual Rainfall.

		inches
Ft. Mohave	10 yrs	2.65
" Yuma..	15 "	3.06
Tulare....	2 "	4.83
San Diego	20 "	9.60
San Jose...	4 "	10.24
Stockton...	9 "	13.25
Monterey...	4 "	15.29
Sta. Barbara	16 "	17.31
Sacramento	29 "	18.76
San Francisco		23..
Humboldt..	11 "	35.02
Colfax....	7 "	42.72
Summit	7 "	58.48
Astoria, Oregon		86...
Marseilles, France		20...
St. Augustine, Fla.		32...
Algiers, Africa		37...
Rome, Italy		39...
St. Louis, Mo		42...
Palatka, Florida		40...
New Orleans, La.		50...
Canton, China		69...

Comparative Temperature of Sea Water.

	Newport, R. I.	Santa Cruz, Cal.	Sta. Barbara, Cal.
Jan	32	52	60
Feb	32	58	61
March...	34	52	61
April....	43	57	61
May.....	52	57	61
June	62	58	62
July.....	66	60	64
Aug	70	60	65
Sept.....	65	60	66
Oct......	58	56	63
Nov	44	55	61
Dec	36	53	60
Mean...	46	56	62

Record for one Year, by L. Bradley, of Aurora, Ill., an Invalid.

Pleasant Days, so that an invalid could be out of doors five or six hours, with safety and comfort	310
Cloudy Days, upon over twenty of which an invalid could be out of doors	29
Showery Days, upon seven of which an invalid could be out of doors an hour at a time several times in each day	12
Windy Days, confining the invalid to the house all day	10
Rainy Days, confining the invalid to the house all day	5

Comparative Mean Temperature of the Six Coldes Months.

Santa Barbara	56.55
City of Mexico	56.03 or .52 colder.
Lisbon, Portugal	54.70 or 1.85 colder.
San Remo, Italy,	53.80 or 2.75 colder.
Mentone, France	53.21 or 3.34 colder.
Nice, Italy	48.45 or 8.10 colder.

Comparative Relative Humidity—Saturation being 100.

While Santa Barbara has not the arid climate of the interior neither has it a humid climate. The adjoining table shows how much dryer the air of Santa Barbara is than that of the other places mentioned.

	Jan'y.	Feb'y.	March	April.	May.	June.	July.	Aug.	Sept.	Oct.	Nov.	Dec.	Year
Santa Barbara	70	68	70	65	61	68	73	72	76	75	65	64	69½
Philadelphia	85½	84	78	70½	75½	78	77	81½	82	79½	80½	85	80
Oakland, Cal.	83	83	77½	85	81	86	86	84	84	86	80	85½	83½
New Orleans	88	84	83	83	84	81	82	87	85	80	84	82	83¼

Record of Temperature at Santa Barbara, Cal.

Latitude, 34° 26'; Longitude, 119° 43'. Height above the Sea, 30 feet.

From January 1, 1883 to January 1, 1884, By Geo. P. Tebbetts.

Day	Jan. 7:00 A.M.	Jan. 12:00 P.M.	Jan. 9:00 P.M.	Feb. 7:00 A.M.	Feb. 12:00 P.M.	Feb. 9:00 P.M.	Mar. 7:00 A.M.	Mar. 12:00 P.M.	Mar. 9:00 P.M.	April 7:00 A.M.	April 12:00 P.M.	April 9:00 P.M.	May 7:00 A.M.	May 12:00 P.M.	May 9:00 P.M.	Jun. 7:00 A.M.	Jun. 12:00 P.M.	Jun. 9:00 P.M.
1	54	40	52	40	58	41	53	59	56	53	62	51	55	65	52	59	68	53
2	52	58	47	41	62	42	53	72	54	40	61	57	54	60	50	52	71	58
3	40	57	40	31	49	33	50	60	53	53	60	50	57	62	53	60	69	70
4	48	57	48	35	40	44	56	64	55	55	61	52	52	65	53	67	93	88
5	43	58	43	40	50	43	47	65	54	51	62	51	55	58	54	74	94	66
6	40	51	44	35	51	40	51	60	53	55	60	53	56	50	52	64	71	63
7	44	60	45	31	53	40	51	57	54	58	61	52	51	60	52	61	71	60
8	45	66	46	30	58	30	53	55	52	53	65	52	55	63	54	61	70	61
9	44	67	43	36	50	41	51	57	52	50	65	52	50	61	55	62	67	62
10	47	70	50	38	56	40	52	60	53	50	65	52	55	57	55	62	67	60
11	51	59	50	40	65	48	53	59	52	52	63	50	50	62	50	60	70	62
12	47	56	40	34	65	43	52	57	54	63	58	40	55	62	55	61	73	63
13	43	60	45	43	55	50	52	60	50	47	64	48	57	66	50	64	74	64
14	45	56	41	[illegible]	54	47	50	62	53	45	60	47	52	62	50	64	67	59
15	40	56	45	43	54	44	51	58	55	54	75	57	55	60	53	62	58	57
16	34	55	40	40	54	44	53	60	54	51	82	55	53	65	55	60	70	60
17	45	60	45	35	57	43	54	58	53	50	63	52	53	65	54	62	72	62
18	39	56	43	44	63	49	55	62	51	51	62	52	56	67	60	61	69	61
19	39	63	40	50	72	55	48	62	53	52	60	50	60	85	67	62	71	60
20	34	53	44	52	65	52	54	62	54	51	59	52	70	92	74	61	72	62
21	41	51	45	49	58	55	51	60	54	53	70	50	80	88	60	65	71	62
22	41	63	44	50	61	53	53	63	53	58	65	68	65	60	58	64	70	62
23	40	63	42	52	59	53	50	62	50	54	63	47	58	68	60	63	78	61
24	41	62	48	50	60	48	53	63	53	44	60	46	54	68	59	64	80	60
25	48	58	52	48	64	48	54	61	50	46	50	50	58	60	50	65	82	63
26	55	67	60	50	72	48	53	62	57	52	62	50	57	70	58	66	89	64
27	41	63	45	50	75	53	55	57	57	50	61	50	56	68	55	67	80	69
28	40	61	43	50	77	57	57	60	57	52	63	53	58	60	57	62	74	60
29	41	61	74	..	..	..	57	62	55	53	58	55	57	60	58	64	70	61
30	72	71	63	..	..	..	50	60	50	51	64	55	58	65	53	63	74	62
31	45	72	54	..	..	..	55	59	48	..	..	..	57	64	54	..	..	..
	44	58	48	43	50	40	53	60	53	52	64	52	57	67	57	63	73	62

Day	July 7:00 A.M.	July 12:00 P.M.	July 9:00 P.M.	Aug. 7:00 A.M.	Aug. 12:00 P.M.	Aug. 9:00 P.M.	Sept. 7:00 A.M.	Sept. 12:00 P.M.	Sept. 9:00 P.M.	Oct. 7:00 A.M.	Oct. 12:00 P.M.	Oct. 9:00 P.M.	Nov. 7:00 A.M.	Nov. 12:00 P.M.	Nov. 9:00 P.M.	Dec. 7:00 A.M.	Dec. 12:00 P.M.	Dec. 9:00 P.M.
1	63	79	60	58	70	60	63	81	66	50	71	65	50	66	53	49	60	56
2	63	73	60	60	70	60	63	81	65	57	75	56	54	67	52	52	62	57
3	62	71	61	58	71	56	67	84	65	54	68	55	48	64	50	57	58	50
4	62	74	65	60	72	54	68	80	63	58	67	54	48	66	46	50	50	50
5	65	70	62	60	70	55	64	69	62	51	68	56	45	65	49	47	59	46
6	62	73	60	60	72	57	63	74	63	58	67	55	50	63	48	45	62	47
7	63	72	61	64	73	61	57	74	63	60	68	55	51	64	50	42	65	50
8	60	72	60	60	71	60	62	75	57	66	70	58	50	62	57	42	66	55
9	62	70	60	60	75	59	62	69	61	48	75	57	50	65	55	49	68	50
10	59	75	64	58	70	60	60	69	62	54	70	53	45	70	49	43	70	48
11	67	76	63	60	69	59	60	68	60	58	67	58	55	74	58	49	70	48
12	62	73	63	62	69	60	60	70	62	58	65	59	53	80	55	43	67	46
13	65	69	59	63	79	61	58	71	60	58	64	59	55	70	49	45	67	47
14	61	70	56	62	76	62	61	72	63	58	62	57	44	65	50	46	66	52
15	60	71	60	61	70	60	62	74	57	57	66	55	50	68	48	51	60	58
16	60	71	61	60	72	60	62	75	60	54	70	53	41	70	48	59	70	58
17	60	73	61	58	69	59	63	71	57	50	63	51	49	61	45	53	67	51
18	60	70	61	58	69	62	60	71	62	49	60	49	53	60	61	53	60	57
19	60	69	61	60	73	62	60	70	61	56	60	49	46	60	54	49	50	56
20	63	73	63	60	70	62	60	72	61	54	68	52	50	62	40	50	56	58
21	64	74	60	60	70	61	59	78	66	56	68	53	44	61	47	57	58	54
22	63	74	59	61	71	60	67	69	71	49	69	57	42	64	46	54	60	65
23	60	72	59	60	68	63	63	84	63	54	60	65	44	65	40	45	50	40
24	60	75	60	59	76	63	60	92	62	60	70	54	43	60	49	40	58	45
25	60	70	61	62	77	62	60	91	60	52	67	57	53	62	48	45	59	50
26	60	65	58	64	75	64	63	73	62	50	64	53	43	69	46	46	65	49
27	60	67	59	61	80	68	61	70	60	49	61	50	44	60	50	42	60	53
28	58	72	60	65	84	66	60	71	60	53	62	50	43	69	60	45	61	46
29	64	79	61	67	82	67	60	70	58	47	76	51	50	67	67	42	50	40
30	65	72	61	64	77	65	64	74	60	49	70	51	49	75	60	45	60	52
31	59	73	60	66	72	62	..	..	..	50	68	54	..	..	..	44	62	48
	64	72	61	61	73	63	60	74	62	51	68	55	48	66	50	48	63	51

Comparative Temperature, At Various Celebrated Health Resorts, and other Places of Note:

Location.	Winter.	Sp'g	Summer.	Autumn	Dif. Sum Win.
Funchal, Madeira	62:88	64:55	70:80	70:19	8:10
St. Michael, Azores	57:83	61:17	68:33	62:33	10:50
SantaCruz, Canaries	64:65	68:87	76:68	74:17	12:03
STA. BARBARA	54:29	59:45	67:71	63:11	13:42
Nassau, Bahama Is.	70:07	77:27	86:00	80:33	15:33
San Diego, California	54:00	60:14	69:67	64:63	15:68
Cadiz, Spain	52:00	59:03	70:43	65:35	17:53
Lisbon, Portugal	53:00	60:00	71:00	62:00	18:00
Malta	57:40	62:70	78:20	71:03	20:74
Algiers	56:00	60:00	77:00	60:00	22:00
St. Augustine, Fl'a	58:25	68:60	80:30	71:90	22:11
Rome, Italy	48:00	57:65	72:16	63:96	23:26
Sacramento, Cal	47:02	59.17	71:19	61:72	23:27
Mentone	49:50	60:00	73:00	59:00	23:50
Nice, Italy	47:88	50:23	72:26	61:03	24:44
New Orleans, La	56:00	69:37	81:08	69:80	25.08
Cairo, Egypt	58:52	73:58	85:10	71:48	26:58
Jacksonville, Fl'da	55:02	68:83	81:93	62:54	26:91
Pau, France	41:80	54:06	70:72	57:39	28:80
Florence, Italy	44:30	56:00	74:00	60:70	29:70
San Antonio, Texas	52:74	70:48	83:73	71:50	30:99
Aiken, S. Carolina	45:82	61:32	77:36	61:90	31:54
Fort Yuma, California	57:96	73:40	92:07	75:06	34:11
Visalia, "	45:38	59:46	80.78	60:34	35:40
Santa Fe, New Mexico	30:28	50:06	70:50	51:34	40:22
Boston, Mass	28:08	45:01	68:68	51:04	40:60
New York, N. Y	31:93	48:26	72:02	48:50	40:60
Albuquerque, N. Mex.	34:78	56:36	76:27	56:33	41:40
Denver, Colorado	27:06	40:33	71:06	47:10	44:00
St. Paul, Minnesota	15:06	41:29	68:03	44:08	52:04
Minneapolis, Minn.	12:87	40:12	68.34	45:33	55:49
.... Means.					

Highest and Lowest Temperature, from Jan. 1, 1871, to Jan. 1, 1884.

		Jan.	Feb.	Mar.	April	May.	June	July	Aug.	Sept.	Oct.	Nov.	Dec.	Year
Highest	1871	70	63	75	83	84	79	89	98	88	100	80	72	100
	1872	72	73	74	75	94	100	82	90	88	88	81	79	100
	1873	76	67	70	80	78	80	86	82	83	82	82	86	86
	1874	70	60	70	78	78	82	86	88	83	72	76	69	88
	1875	73	77	70	80	95	80	84	90	83	72	78	69	95
	1876	73	71	74	80	80	84	84	89	83	85	81	75	89
	1877	83	79	76	79	75	102	87	82	86	77	82	74	102
	1878	67	69	69	77	77	70	80	84	94	84	70	76	94
	1879	68	79	89	83	92	97	80	85	86	90	77	71	97
	1880	70	69	68	69	82	75	75	77	85	81	77	70	85
	1881	66	78	70	70	74	78	87	84	80	73	73	72	87
	1882	62	68	78	72	78	77	80	81	83	82	75	80	83
	1883	72	77	72	82	92	92	79	84	98	75	90	70	98
Lowest	1871	40	40	40	40	52	56	58	60	54	48	40	40	40
	1872	38	42	46	37	50	57	60	60	58	49	51	41	33
	1873	42	40	48	46	54	58	58	61	50	50	46	40	40
	1874	38	41	44	50	53	56	58	58	57	52	48	43	38
	1875	40	44	44	41	50	54	58	59	57	52	48	43	40
	1876	40	41	40	47	53	56	59	60	56	55	40	43	40
	1877	42	46	46	50	52	58	61	58	58	46	49	40	40
	1878	38	41	46	46	52	57	58	57	50	52	41	37	37
	1879	38	39	40	50	52	54	58	50	55	50	42	33	33
	1880	31	36	39	40	51	52	53	54	51	41	30	41	30
	1881	40	43	43	40	52	55	57	56	53	45	36	39	36
	1882	33	33	42	43	50	53	52	58	50	47	38	38	33
	1883	39	31	47	44	50	52	58	56	58	47	41	40	31

Average for January, 53.25. Average for July, 68.45. Yearly Average, 61.43. Difference between July and January, 15 degrees.

No. of Days the Temperature got below 42° or above 82°.

	1873	1874	1875	1876	1877	1878	1879	1880	1881	1882	1883	Average.
Days below 42 degrees	2	0	2	10	6	12	18	24	12	41	30	15
Days above 82 degrees	2	13	12	7	13	10	12	2	2	3	12	8

RAINFALL AT SANTA BARBARA, From Records of Dr. Shaw and G. P. Tebbetts.

Year.	68-69	69-70	70-71	71-72	72-73	73-74	74-75	75-76	76-77	77-78	78-79	79-80	80-81	81-82	82-83	83-84	Mean
Oct.	....	.30	1.04	.09	....	....	1.01	...	....	.....	.36	.44	.17	1.09	.37	1.35	.48
Nov.	1.25	.06	.27	1.83	....	.27	1.30	6.53	....	1.41	...	1.93	.20	.40	.77	....	1.05
Dec.	4.20	.57	1.41	6.56	4.34	5.30	...	.81	...	3.65	0.89	5.01	0.94	.96	.10	2.80	3.25
Jan.	3.26	.25	.80	2.53	.58	4.54	14.84	7.50	3.04	7.87	4.83	1.42	3.02	1.30	2.18	6.54	4.06
Feb.	2.12	5.37	2.02	1.81	6.48	3.17	.18	5.07	....	12.32	.72	11.50	.30	2.85	2.05	10.97	4.30
Mar.	4.22	.83	.02	.13	.06	.78	.38	2.73	.01	2.08	.34	1.22	1.30	0.85	3.27	8.02	2.15
April	.46	.09	2.02	1.80	....	.28	.10	.27	.39	3.34	1.80	6.25	.63	2.82	.33	2.02	1.51
May.	.20	.74	.37	....	....	.14	...	....	.46	.20	.30	....	....	....	2.05	.40	.35
June	....	.07	....	.14	....	.....	....	....	....	.06	.11	....	....	....	.35	1.75	.15
July	....	....	....	....	....	....	....	....	...	....	....	.07	....	.06	....	....	.01
Aug.	....	....	....	....	....	....	....	....	....	....	....	....	....	....	....	.....	.....
Sept	....	....	....	....	....	....	....	....	....	....	....	.11	....	....	....	....	.01
Total	15.77	10.27	8.01	14.04	10.45	14.44	18.71	21.07	4.49	31.51	15.34	27.96	15.68	10.93	12.97	35.53	16.32

Average annual rainfall for fifteen years, 17.31 inches.

Comparative Annual Rainfall.

		inches
Ft. Mohave	10 yrs	2.05
" Yuma	15 "	3.06
Tulare	9 "	4.83
San Diego	20 "	9.90
San Jose	4 "	10.24
Stockton	9 "	13.25
Monterey	4 "	15.29
Sta. Barbara	16 "	17.31
Sacramento	28 "	18.76
San Francisco		21.
Humboldt	11 "	35.09
Colfax	7 "	42.72
Summit	7 "	56.46
Astoria, Oregon		86.
Marseilles, France		20.
St. Augustine, Fla.		32.
Algiers, Africa		37.
Rome, Italy		30.
St. Louis, Mo		42.
Palatka, Florida		49.
New Orleans, La.		50.
Canton, China		69.

Comparative Temperature of Sea Water.

	Jan	Feb	March	April	May	June	July	Aug	Sept.	Oct.	Nov	Dec	Mean
Newport, R. I.	32	32	34	43	52	62	69	70	63	54	44	36	48
Santa Cruz, Cal.	52	58	52	57	57	58	60	60	60	56	55	53	56
Sta. Barbara, Cal.	60	61	61	61	61	62	64	65	68	63	61	60	62

Record for one Year, by L. Bradley, of Aurora, Ill., an Invalid.

Pleasant Days, so that an invalid could be out of doors five or six hours, with safety and comfort	310
Cloudy Days, upon over twenty of which an invalid could be out of doors	20
Showery Days, upon seven of which an invalid could be out of doors an hour at a time several times in each day	12
Windy Days, confining the invalid to the house all day	10
Rainy Days, confining the invalid to the house all day	5

Comparative Mean Temperature of the Six Coldest Months.

Santa Barbara	56.55
City of Mexico	56.03 or .52 colder.
Lisbon, Portugal	54.70 or 1.85 colder.
San Remo, Italy	53.80 or 2.75 colder.
Mentone, France	53.21 or 3.34 colder.
Nice, Italy	48.45 or 8.10 colder.

Comparative Relative Humidity—Saturation being 100.

While Santa Barbara has not the arid climate of the interior neither has it a humid climate. The adjoining table shows how much dryer the air of Santa Barbara is than that of the other places mentioned.

	Jan'y.	Feb'y.	March	April.	May.	June.	July.	Aug.	Sept.	Oct.	Nov.	Dec.	Year
Santa Barbara	70	68	76	65	61	68	73	72	70	75	65	64	69½
Philadelphia	85½	84	78	70½	75½	78	77	81½	82	70½	80½	85	80
Oakland, Cal	83	83	77½	85	81	80	86	84	84	88	80	85½	83½
New Orleans	88	84	83	83	84	81	82	87	85	80	84	82	83½

www.ingramcontent.com/pod-product-compliance
Lightning Source LLC
Chambersburg PA
CBHW021943190326
41519CB00009B/1129
9783337379681